KW-052-560

# CIBS CODE

# CODE FOR
# INTERIOR LIGHTING
# 1984

**CIBS**

© The Chartered Institution of Building Services
Delta House, 222 Balham High Road, London SW12 9BS

The rights of publication or of translation are reserved.
No part of this publication may be reproduced, stored in a retrieval system or
transmitted in any form or by any means without the prior permission of
the Institution.

ISBN 0 900953 27 6

1936 *First Published*
1938 *Revised*
1941 *Revised*
1942 *Revised*
1944 *Revised*
1945 *Revised*
1946 *Revised*
1949 *Revised*
1955 *Revised*
1961 *Revised*
1968 *Revised*
1973 *Revised*
1977 *Revised*
1984 *Revised*
1985 *Reprinted with amendments*

©
THE CHARTERED INSTITUTION OF BUILDING SERVICES
LONDON
1984

*Printed in Gt. Britain by Yale Press Ltd., London SE25 5LY*

# Foreword

In 1980 the Lighting Division Technical Committee of the Chartered Institution of Building Services appointed a Task Group to revise the 1977 Code for Interior Lighting, published by the Illuminating Engineering Society.

Extensive enquiries were made with users to ensure that the categories of tasks used in the recommendations were still valid and useful. The replies received from Members set the aims of this edition, and expanded its scope.

The Institution is indebted to the various government and local authority departments, professional and trade associations, academic institutions and individuals in the UK and overseas for their constructive comments on the previous editions and drafts of this document.

**Task Group**

| | |
|---|---|
| P.R. Boyce *(Chairman)* | D.L. Loe |
| J.E. Baker | J.A. Lynes |
| R.I. Bell | P.C.M. McCarthy |
| W. Carlton | P.G.T. Owens |
| M.B. Clark | D.J. Paine |
| J.B. Collins | D. Poole |
| P.E. Donnachie | W.A. Price |
| A.E. Fothergill | J.C. Procter |

| *Technical Secretary* | *Co-ordinating Editor* |
|---|---|
| K.J. Butcher | V.P. Rolfe |

# Contents

# Part 3    Equipment

# Part 4    Lighting Design

# Appendices

# Preface

Lighting affects almost every aspect of our lives. It is used in places as diverse as factories, offices, libraries, restaurants, schools and shops, to produce conditions which enable us to see what we are doing, and to create an ambience appropriate to the setting. This Code has been prepared with the aims of: *(a)* specifying the lighting conditions appropriate for a wide range of interiors, and *(b)* offering guidance on methods of obtaining those conditions. It is organised in four parts. Part 1 summarises the effect of lighting conditions on the performance of tasks, the appearance of the interior and the comfort of the occupants. Part 2 contains recommendations of lighting conditions suitable for a large number of applications. Part 3 describes the properties of lighting equipment available commercially. Part 4 sets out suitable design procedures.

Lighting is both an art and a science. It can be both decorative and functional although the balance between decoration and function will vary with application. The applications considered in this Code are essentially functional. Even so, given the differences in visual work, building form and surface finish that occur, and the different light sources and luminaires that are available, there is plenty of opportunity for variety in design. For this reason the recommendations made in this Code are given in terms of the end result rather than the means of producing it. The route by which the end results are achieved is determined by the designer.

The recommendations given in this Code are representative of good practice. They are the result of considering scientific knowledge, practical experience, technical feasibility and economic reality. The recommendations have no statutory standing, although some may be adopted by appropriate authorities. Taken together, the recommendations represent a base for designers to build on. They offer guidance rather than demand compliance.

## Part 1

The visual effects of lighting

Fig 1.1   Daylighting

Fig 1.2   Electric lighting/daylighting

Fig 1.3   Electric lighting

## 1.1  Introduction

The lighting of an interior should fulfill three functions. It should: (a) ensure the safety of people in the interior; (b) facilitate the performance of visual tasks; (c) aid the creation of an appropriate visual environment.

Safety is always important but the emphasis given to task performance and the appearance of the interior will depend on the nature of the interior. For example, the lighting considered suitable for a factory toolroom will place much more emphasis on lighting the task than on the appearance of the room, but in a hotel lounge the priorities will be reversed. This variation in emphasis should not be taken to imply that either task performance or visual appearance can be completely neglected. In almost all situations the designer should give consideration to both these aspects of lighting.

Lighting affects safety, task performance and the visual environment by changing the extent to, and the manner in which different elements of the interior are revealed. Safety is ensured by making any hazards visible. Task performance is facilitated by making the relevant details of the task easy to see. Different visual environments can be created by changing the relative emphasis given to the various objects and surfaces in an interior. Different aspects of lighting influence the appearance of the elements in an interior in different ways. This part of the Code discusses the influence of each important aspect of lighting separately. However, it should always be remembered that lighting design involves integrating the various aspects of lighting into a unity appropriate to the design objectives. This process is discussed in Part 4.

## 1.2  Daylight and electric light

One of the most fundamental decisions to be made when designing the lighting of an interior is the relationship between daylight and electric light. This can take three forms (a) to rely on daylight during daytime and to design the electric lighting only for night-time conditions (Fig 1.1) (b) to use daylight as available but supplement it as required by electric lighting (Fig 1.2), and (c) to ignore daylight and operate the building on electric lighting only (Fig 1.3). The decision as to which of these relationships to adopt will be influenced by many considerations in addition to the lighting effects. For example, the energy consumption and costs involved, the possible building forms, and the need for a controlled environment are all relevant factors in determining the relationship between daylight and electric light. Nonetheless the roles assigned to daylight and electric light do change the lighting conditions produced so it is appropriate to indicate some of the facts which should be considered by the designer. There is little doubt that given a choice people prefer to work by daylight and to enjoy a view. Windowless interiors are generally disliked, particularly if they are small, but may be accepted if there are good reasons for the interior to be windowless. However, people do not like the uncomfortable thermal conditions which extensive daylighting can produce, so it is hardly surprising that the most common approach is where daylight and electric light are combined to produce sufficient and suitable lighting on the task and in the room, by day and night. Then, the electric lighting serves to supplement daylight when and where it is insufficient and the daylight contributes an element of variation and directional flow to the appearance of the interior.

# 1.3 Lighting levels

The lighting level produced by a lighting installation is usually quantified by the illuminance produced on a specified plane. In most cases this plane is the major plane of the tasks in the interior, and is commonly called the working plane. The illuminance provided by an installation affects both the performance of the tasks and the appearance of the space.

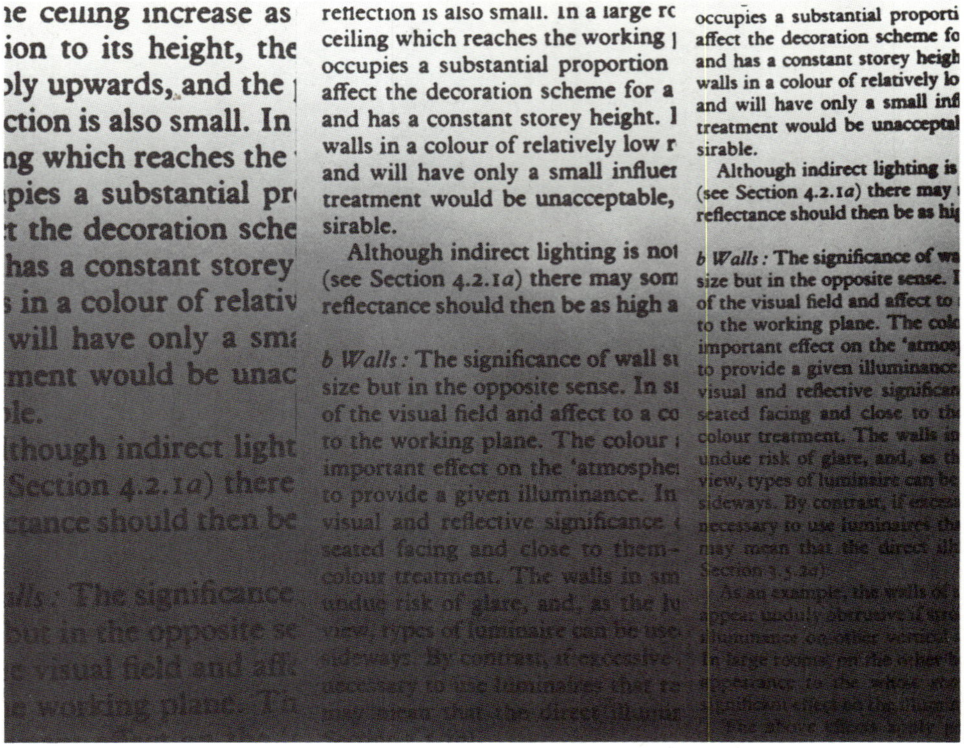

Fig 1.4 The effect of varying size and contrast of task details on ease of reading

## 1.3.1 Task performance

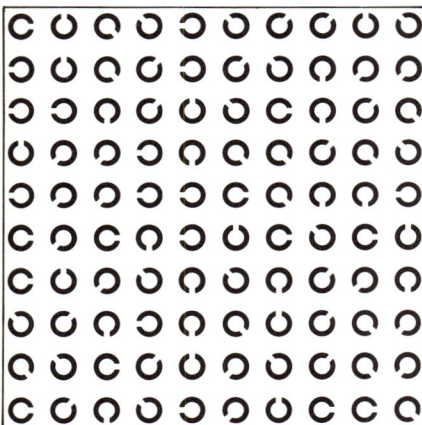

Fig 1.5 Landolt rings

The effect of lighting on work depends on the size of the critical details of the task and on their contrast with their background (Fig 1.4). One particular task much used for laboratory investigations of visual performance consists of scanning an array of Landolt rings (Fig 1.5) and identifying those rings with gaps in a specified direction. By changing the size of the gap and the contrast of the ring with the background it is possible to vary the difficulty of the task over a wide range.

Figure 1.6 shows the effect of illuminance on the performance of the laboratory task illustrated in figure 1.5. Three important points should be noted from figure 1.6. The first is that increasing the illuminance on the task produces an increase in performance following a law of diminishing returns. The second is that the illuminance at which performance levels off is dependent on the visual difficulty of the task, i.e. the smaller the size and the less the contrast of the task the higher the illuminance at which performance saturates. The third is that although increasing illuminance can increase task performance, it is not possible to bring a difficult visual task to the same level of performance as an easy visual task simply by increasing the illuminance.

In principle these effects occur for all tasks, although the exact relationship between the illuminance on the task and the performance achieved will vary with the nature of the task. There are two aspects of a task which are important in determining the effect of illuminance. One has already been mentioned: the task difficulty; the greater the visual difficulty the greater the importance of illuminance. The other is the extent to which the visual part of the task determines the overall performance. Where there is only a small visual component, as in audio typing, the influence of illuminance on overall task performance is likely to be small but where the visual component is a major element of the complete task, as in copy typing, then the illuminance provided will be important.

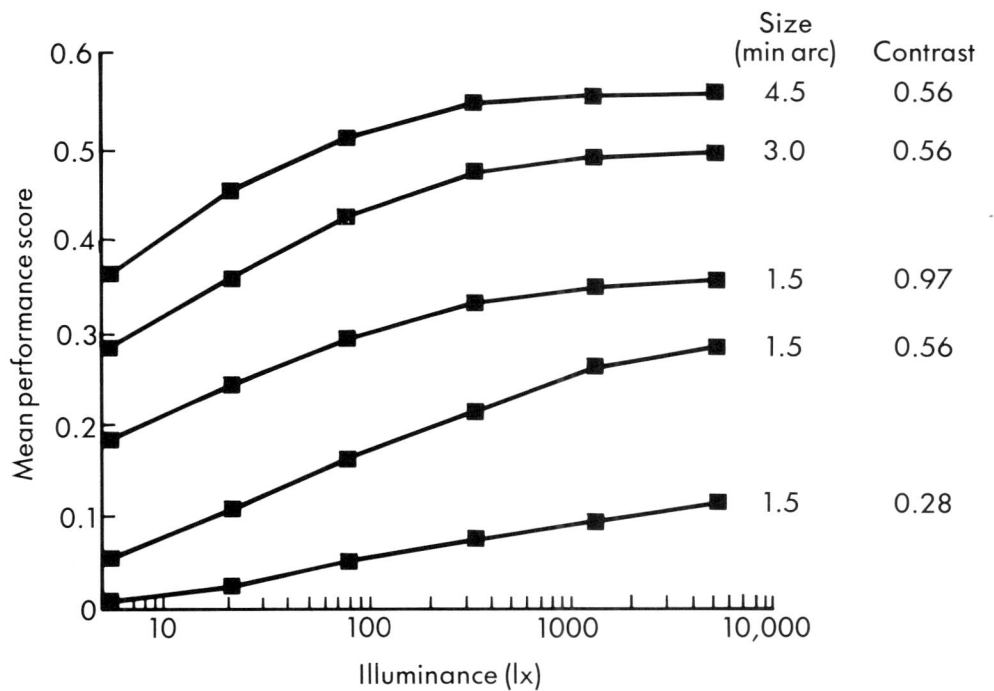

**Fig 1.6** Mean performance scores for Landolt ring charts (after Weston, H.C., Industrial Health Research Board, Report 87, London, HMSO, 1945)

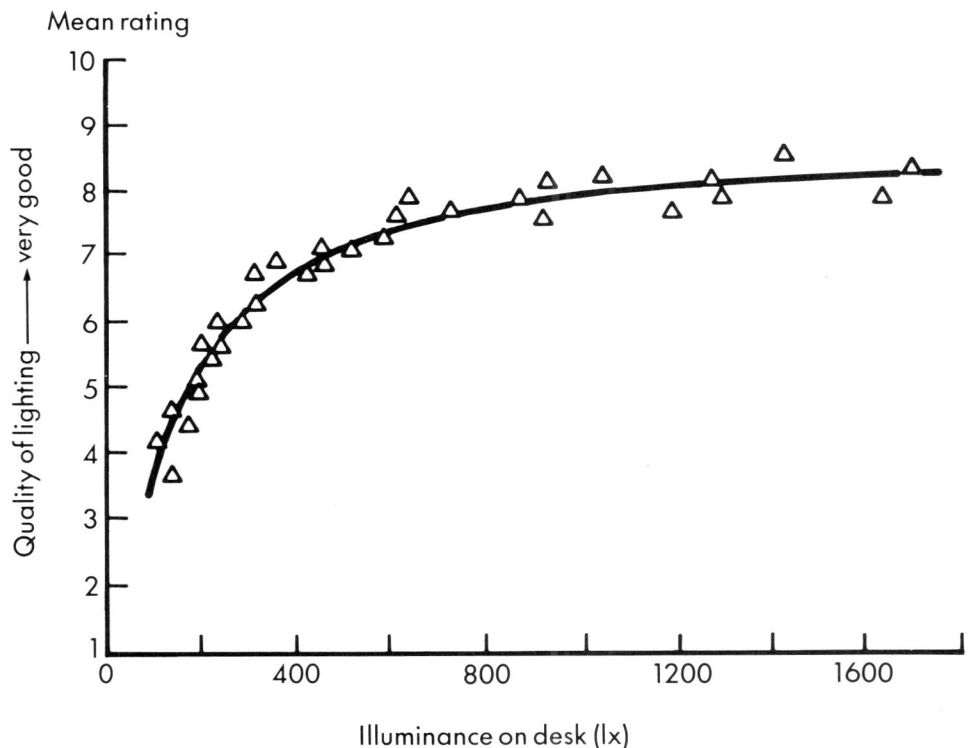

**Fig 1.7** Mean assessment of quality of lighting (after Saunders, J.E., L.R. and T., 1(1), 39 (1969))

## 1.3.2  Preference

Figure 1.7 shows mean assessments of the quality of lighting obtained in an office lit uniformly by a regular array of luminaires. Increasing the illuminance on the plane of the desk increases the perceived quality of the lighting until it saturates at about 800 lx. This demonstrates the importance of the illuminance as one factor in determining people's satisfaction with an interior.

There is no sharp cut off where lighting conditions move from the dreadful to the wonderful. Figure 1.7, like figure 1.6 shows that as illuminance increases from a low level there is initially a rapid improvement, but as illuminance continues to increase, the improvement becomes smaller, until eventually it ceases altogether. So identifying a suitable illuminance for an interior is inevitably a matter of judgement. Recommended illuminances for specific applications are given in Part 2. However, there is one generally applicable recommendation. It is that no continuously occupied working space should have an illuminance of less than 200 lx on the working plane.

### 1.3.3 Alternative measures

**Fig 1.8** Not all working planes are horizontal

The recommended illuminance given in Part 2 should always be provided on an appropriate plane. This plane can be horizontal, vertical or inclined, depending on the situation. Figure 1.8 shows the results of providing the recommended illuminance on the wrong plane. But even when an appropriate plane is used the illuminances on other planes should not be ignored. In rooms where the surface reflectances are low and/or the distribution of light is strongly directional, the illuminance on a single plane can give a misleading indication of the appearance of the room. For such situations the quantity of light is more usefully described by measures of either scalar illuminance or mean cylindrical illuminance (see appendix 1).

### 1.3.4 Uniformity of illuminance

Uniformity of illuminance can be considered over two areas: on and around the task itself, and over the whole interior. For the task area and its immediate surround, uniformity of illuminance is important. Sudden changes in illuminance in this region are likely to cause distraction and dissatisfaction and may affect task performance. The ratio of the minimum illuminance to the average illuminance, over the task area, should not be less than 0.8. A greater degree of non-uniformity is acceptable between the illuminance on the task area and the parts of the room which do not contain tasks, although a minimum uniformity is still necessary to avoid discomfort. The illuminance on areas adjacent to the task area should not be less than one-third of the task illuminance.

## 1.4 Directional effects

The directional distribution of light in a space is important to the appearance of objects and consequently for task performance and the perception of the space. The effect of directional lighting on the appearance of objects depends on the form and the surface characteristics of the objects. Figure 1.9 shows three different objects standing in the same incident illumination. All show the pattern expected from the variation in illuminance over their surfaces but, in addition, the glossy object shows highlights and the sculptured object has a marked shadow pattern. The effect of the directional distribution of light on an object can be described in terms of the illuminance pattern, the highlight pattern and the shadow pattern, but no complete description of the way in which lighting affects the appearance of objects has yet been developed.

**Fig 1.9** The 'flow of light' over three objects

## 1.4.1 Revealing form

The strength of directional lighting at a point can be quantitified by the ratio of the magnitude of the illumination vector to the scalar illuminance (appendix 1). This quantity is known as the vector/scalar ratio. The direction of the flow of light is given by the direction of the illumination vector. Figure 1.10 (see over) displays the effect of different vector/scalar ratios and different illumination vector directions on the appearance of people's faces and describes some typical lighting conditions in which these appearances are produced. No single value of vector/scalar ratio is right for all purposes but, for general use where the perception of faces is important, vector/scalar ratios in the range of 1.2-1.8 will be satisfactory.

There is some evidence that directions of the illumination vector in the range 15° to 45° from the horizontal are preferred. This condition can be readily achieved in rooms lit during daytime by side windows but is very difficult to achieve at night when only electric light is in use. For practical reasons, most electric lighting is ceiling mounted, so the vector direction is almost always vertically downward. If, for this situation, the vector/scalar ratio is high, harsh and unnatural shadows will be produced on the face. To overcome this situation the designer has to rely on light reflected from the room surfaces to soften the shadow.

## 1.4.2 Revealing texture

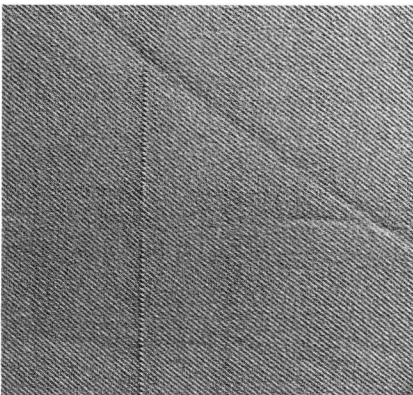

**Fig 1.11** Directional lighting indicating a pulled thread in material

The extent to which texture is revealed by a lighting installation is determined by the angle of incidence of its illumination vector, regardless of source size. The revealing of texture can have a marked effect on the appearance of a space; it can also be important for visual inspection work (see appendix 2). Figure 1.11 shows the effect of lighting at glancing angles on the ease of seeing a pulled thread in a piece of material. In this case the shadows and highlights created around the thread are beneficial, but shadows and highlights can reveal too much texture which can be detrimental to task visibility. For this reason, any directional effect which is used to enhance task visibility should be 'tailored' to the application.

**Figure 1.10   Relationship of vector/scalar ratio to assessment of directional qualities of the lighting\***

| Vector direction between 15° and 40° below horizontal | Vector direction predominantly vertically downwards | Vector/scalar ratio and directional strength | Typical assessment of directional qualities | Typical conditions |
|---|---|---|---|---|
| 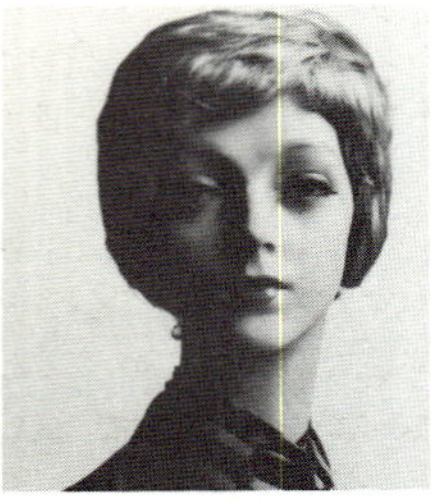 |  | 3·0<br><br>Very strong | Strong contrasts; detail in shadow not discernible | Selective spotlighting; direct sunlight |
| 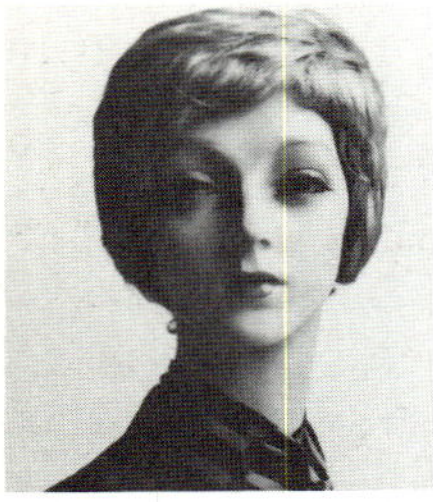 |  | 2·5<br><br>Strong | Noticeably strong directional effect; suitable for display, generally too harsh for human features | Luminaries of narrow light distribution, low flux fraction ratio, dark floor. Windows on one side, dark surfaces |
| 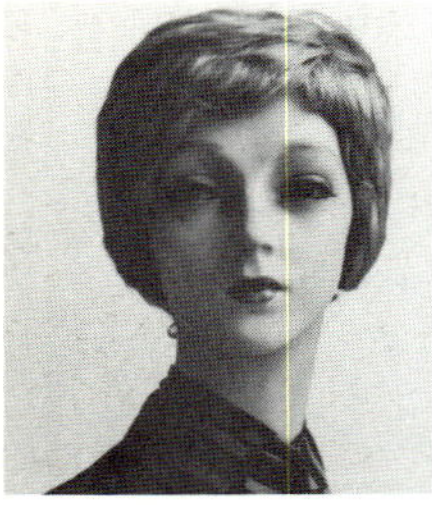 | 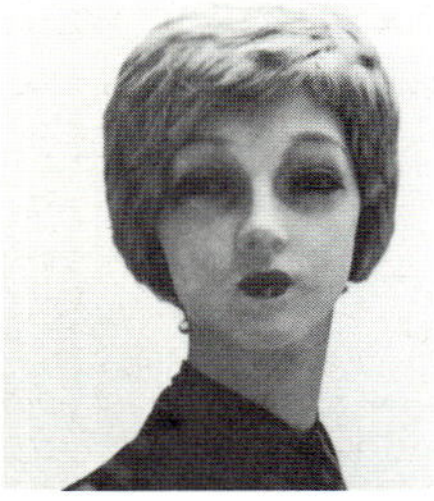 | 2·0<br><br>Moderately strong | Pleasant appearance of human features for formal or distant communication | Luminaires of narrow light distribution with medium or light floor. Luminaires of medium or wide light distribution with dark floor. Side windows with light room surfaces. |
|  |  | 1·5<br><br>Moderately weak | Pleasant appearance of human features for informal or close communication | |
| 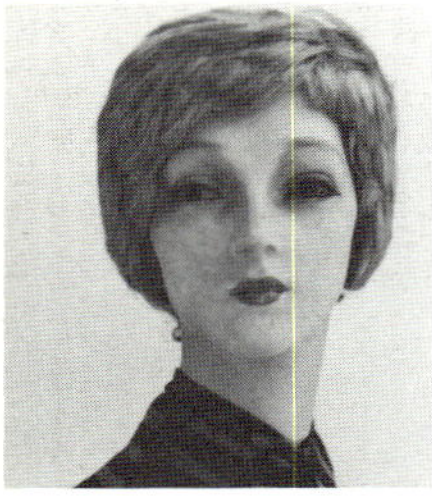 |  | 1·0<br><br>Weak | Soft lighting effect for subdued contrasts | Luminaires of medium or wide light distribution with light floor. Side windows in opposite walls |
| 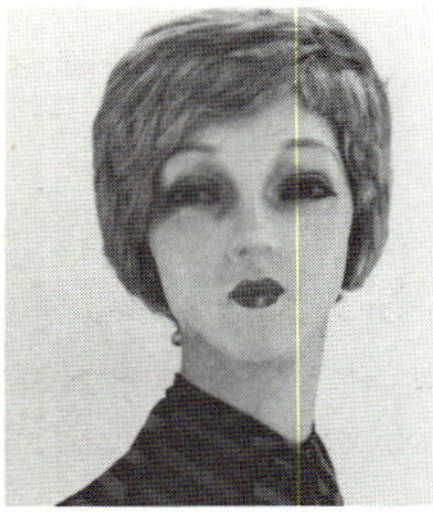 |  | 0·5<br><br>Very weak | Flat, 'shadow-free' lighting; directional effect not discernible | Luminous ceiling or indirect lighting with light room surfaces |

\*See Appendix 11 for methods of measurement of vector/scalar ratio and section 4.4.3.3 for a means of predicting average vector/scalar ratio for a regular array of luminaires.

# 1.5 Surfaces

The effect a lighting installation creates in an interior is strongly influenced by the properties of the room surfaces. For this reason, if for no other, the lighting designer should always attempt to identify the proposed surface finishes early in the design process. The main properties of the room surfaces which are relevant to the appearance of the space are their reflectance and their colour.

## 1.5.1 Surface reflectances

For interiors lit from the ceiling, the significance of the ceiling reflectance increases as the room area increases. In small rooms the ceiling is not conspicuous and its contribution to the illuminance on the working plane is usually small. In a large room the contribution of light reflected from the ceiling to the total illuminance on the working plane is usually large and the ceiling occupies a substantial proportion of the visual field. For large rooms a white or near white ceiling is desirable. The reflectance of the ceiling cavity should be at least 0.6; this usually requires a surface reflectance of at least 0.8. In small rooms a low reflectance ceiling may be acceptable, although if the room is predominantly lit by daylight from side windows the room may appear gloomy if too low a reflectance is chosen. Where indirect lighting is used a white or near white ceiling is desirable, regardless of room size.

Wall reflectance is usually unimportant to the lighting of a large room except for the area immediately adjacent to the wall. If low wall reflectances are used the illuminance in the adjacent areas may be too low. In small rooms wall reflectance is always important. High wall reflectances will enhance the illuminance on the working plane and increase the inter-reflective component of the lighting, thereby increasing uniformity. The importance of having a high wall reflectance is increased when the room is predominantly lit by daylight from side windows. In all rooms, unless a high reflectance finish is applied to the window wall, the luminance difference between the window wall and the view through the window may be excessive and uncomfortable.

All this suggests that a high reflectance finish to walls is highly desirable. However, the use of high reflectance wall finishes should be treated with caution. Large areas of high reflectance may compete for attention with the task areas and may lead to eyestrain and feelings of discomfort. Further, if the high reflectance surfaces are produced by gloss paint, reflected glare is likely to occur. In working interiors, the effective reflectance of the principal walls should be between 0.3 and 0.7. This usually means that the wall surface finish should have an actual reflectance greater than 0.5 to satisfy the lower figure. The upper figure will only be reached with light surface finishes on a wall free of windows, darker hangings or furniture. The reflectance of window wall surfaces (including reveals and glazing bars but excluding glazing) should be at least 0.6 to reduce contrast with the bright scene outdoors during daytime. Where the perception of people's faces is particularly important, e.g. lecture theatres, conference rooms, the reflectance of walls which form the background against which people are seen should not exceed 0.6.

Dark floor cavities will tend to make ceilings and walls look underlit, especially when daylight from side windows is used, but using very light floors tends to create a maintenance problem. Where practicable, floor cavity reflectance should be within the range 0.2 to 0.3. This generally involves choosing reflectances of not less than 0.3 for the relevant surfaces.

## 1.5.2 Surface colours

Surface colour can be classified by the use of the Munsell system or by BS 5252: Framework for colour co-ordination for building purposes (see appendix 3). In the Munsell system, each colour is specified by three quantities, its Hue, its Value, and its Chroma. Hue describes whether a

**Fig 1.12** Use of colour to enhance the appearance of an interior

**Fig 1.13** Surfaces of different Chroma used to create a pattern of emphasis (illustration courtesy of Concord-Rotaflex Ltd)

**Fig 1.14** Surfaces of different Hue used to create an 'active' atmosphere

### 1.5.3 Object colours

colour is basically red, yellow, green, blue or purple, etc. Value describes the lightness of the colour and is related to its reflectance. Chroma describes the strength of the colour. This classification forms a convenient basis on which to discuss the effects of room surface colour on the appearance of space (Fig 1.12). By choosing different Values for different components of the interior it is possible to dramatise or to buffer the pattern of light and shade created by the lighting. An example of this is the use of a high reflectance (high Value) wall opposite a window wall.

By choosing colours of different Chromas it is possible to create a pattern of emphasis (Fig 1.13). Strong emphasis requires strong Chromas but their use calls for caution. An area of awkward shape which might pass unnoticed at weak Chroma can looked unsightly at strong Chroma. Also a small area of strong Chroma might be stimulating but the same Chroma over a large area could be overpowering.

The selection of Hue is partly a matter of fashion and partly a matter of emotion. By choosing a predominant Hue for a space it is possible to create a 'cool' or a 'warm', a 'restful' or an 'active atmosphere (Fig 1.14).

All rooms will have a mixture of colours. This fact raises the question of colour harmony. There are a number of so called rules of colour harmony which have little basis in fact. However it is widely believed that the main variable influencing pleasant colour harmonies is the difference in Value for the two colours compared; the greater the difference in Value the greater the chances of achieving a pleasant colour combination. The effect of Chroma differences is thought to be similar, combinations of colours with large differences in Chroma tending to be pleasant. As for Hue differences, there is not believed to be any consistent effect, all the same Hues closely related Hues, or complementary Hues, being capable of creating either pleasant or unpleasant colour combinations. These observations suggest that when selecting colours for an interior the first aspect to consider is the Value of the colours, then the Chroma and finally the Hue. However, once the pattern of light, shade and emphasis has been established by the choice of the Value and Chroma for different surfaces, the range of Hues that are available may be limited. For example, if a given surface is to have both strong Chroma and high Value, then it must inevitably have a yellowish Hue. Conversely, when a surface is required to have low Value and strong Chroma, inevitably a colour from the red to blue part of the Hue circle must be used. Once the level of Chroma is reduced from a high level the whole range of colours is available. It should be apparent from this that the approach suggested for selecting colours is basically a method of ordering thoughts.

One thought that should always be borne in mind is the effect of inter-reflected light. The light reflected from a surface of strong Chroma will be coloured and may influence the colour of other surfaces. The most common situation where this is seen is when a floor covering of strong Chroma is lit by a lighting installation which does not light the ceiling directly. In this situation, the ceiling will mainly be lit by light reflected from the floor and may appear to be a similar colour to the floor.

The colours of objects within an interior can have a marked effect on the appearance of the space. In choosing a combination of colours for both the surfaces and equipment within a space it is preferable if the elements can be considered as a whole so that a degree of visual coordination can be achieved. The actual choice of a combination of colours to produce a coordinated colour scheme is probably one of the most elusive design tasks and at present there is no single widely accepted design procedure.

There are limitations to the choice of colours of some objects within the space. These arise from the use of colour for the coding of services and to indicate potential hazards. The use of colour for the coding of services is governed by BS 1710 and should be undertaken sparingly with emphasis

*Surfaces*

given to identification of outlets, junctions and valves. The use of colour to identify potential hazards is governed by BS 5378. Care should be taken to avoid confusion between BS 5378 hazard warning colours, BS 1710 service colours and other colours in the interior. Care is also necessary to ensure that the lighting does not distort the colours reserved for services or hazard indication in such a way as to be confusing.

# 1.6 Light source colour

Light sources, both natural and electric, have two colour properties related to the spectral composition of their emission. One is the apparent colour of the light that the source emits and the other is the effect that the light has on the colours of surfaces. The latter effect is called colour rendering.

### 1.6.1 Apparent colour of the emitted light

The colour of the light emitted by a 'near white' source can be indicated by its correlated colour temperature (CCT). Each lamp type has a specific correlated colour temperature but for practical use, the correlated colour temperatures have been grouped into three classes by the Commission International de l'Eclairage (CIE) (table 1.1).

**Table 1.1 Correlated colour temperature classes and colour rendering groups used in this Code**

| Correlated Colour Temperature (CCT) | CCT Class |
|---|---|
| $CCT \leqslant 3300K$ | Warm |
| $3300K < CCT \leqslant 5300K$ | Intermediate* |
| $5300K < CCT$ | Cold |

| Colour rendering groups | CIE general colour rendering index ($R_a$) | Typical application |
|---|---|---|
| 1A | $R_a \geqslant 90$ | Wherever accurate colour matching is required, e.g. colour printing inspection. |
| 1B | $80 \leqslant R_a < 90$ | Wherever accurate colour judgements are necessary and/or good colour rendering is required for reasons of appearance, e.g. shops and other commercial premises. |
| 2 | $60 \leqslant R_a < 80$ | Wherever moderate colour rendering is required. |
| 3 | $40 \leqslant R_a < 60$ | Wherever colour rendering is of little significance but marked distortion of colour is unacceptable. |
| 4 | $20 \leqslant R_a < 40$ | Wherever colour rendering is of no importance at all and marked distortion of colour is acceptable. |

*This class covers a large range of correlated colour temperatures. Experience in the U.K. suggests that light sources with correlated colour temperatures approaching the 5300K end of the range will usually be considered to have a 'cool' colour appearance.*

The choice of an appropriate apparent colour of light source for a room is largely determined by the function of the room. This may involve such psychological aspects of colour as the impression given of warmth, relaxation, clarity, etc., and more mundane considerations such as the need to have a colour appearance compatible with daylight and yet to provide a 'white' colour at night. The only general rules to help with the selection of apparent colour are *(a)* for rooms lit to an illuminance of 300 lx or less, a warm or intermediate colour is preferred; cold apparent colour lamps tending to give rooms a gloomy appearance at such illuminances and *(b)* different apparent colour lamps should not be used haphazardly in the same room.

### 1.6.2 Colour rendering

The ability of a light source to render colours of surfaces accurately can be conveniently quantified by the CIE general colour rendering index. This index is based on the accuracy with which a set of test colours are reproduced by the lamp of interest relative to how they are reproduced by an appropriate standard light source, perfect agreement being given a value of 100. The CIE general colour rendering index has some limitations but it is the most widely accepted measure of the colour rendering properties of light sources (see appendix 4). Table 1.1 shows the groups of the CIE general colour rendering index used by the CIE and in this Code.

Where work involving accurate colour judgement is to be done, electric light sources with high CIE general colour rendering indices (i.e. from Groups 1A or 1B) are necessary. Where exact colour matching is to be done, lamps of colour rendering group 1A should be used and the recommendations of BS 950 should be followed as appropriate. The surfaces of surrounding areas where accurate colour judgements are being made should be of weak Chroma (not greater than 1) and medium reflectance (not less than 0.4). An illuminance of at least 500 lx should be provided on the task.

Where the main consideration is the appearance of the space and objects within it, light sources with a high CIE general colour rendering index may be desirable. In general, light sources with good colour rendering properties (Groups 1A and 1B) make surfaces of objects appear more colourful than do light sources with moderate or poor colour rendering properties (Groups 2, 3 and 4). In addition, light sources with poor colour rendering properties may distort some colours to a marked extent. Thus, where a colourful appearance is desirable, lamps with good colour rendering properties are appropriate. However, the exact level of colour rendering desirable in any particular circumstance remains a matter of individual judgement. Ultimately the CIE general colour rendering index is no substitute for actually seeing the effect of different light sources when it comes to assessing their contribution to the appearance of an interior.

# 1.7 Glare

Glare occurs whenever one part of an interior is much brighter than the general brightness in the interior. The most common sources of excessive brightness are luminaires and windows, seen directly or by reflection. Glare can have two effects. It can impair vision, in which case it is called disability glare (Fig 1.15) and it can cause discomfort, in which case it is called discomfort glare (Fig 1.16). Disability glare and discomfort glare can occur simultaneously or separately.

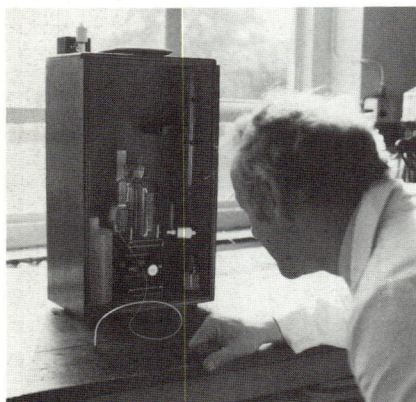

**Fig 1.15** Disability glare from bright sky

### 1.7.1 Disability glare

Disability glare is most likely to occur when there is an area close to the line of sight which has a much higher luminance than the object of regard. Then, scattering of light in the eye and changes in local adaptation can cause a reduction in the contrast of the object. This reduction in contrast may be sufficient to make important details invisible and hence may influence task performance. Alternatively, if the source of high luminance is viewed directly, noticeable after-images may be created. The most common sources of disability glare indoors are the sun and/or sky seen through windows (Fig 1.15) and electric light sources seen by reflection (Fig 1.17). Care should be taken to avoid disability glare in interiors by providing some method of screening windows and avoiding the use of highly specular surfaces. A formula which can be used to estimate the effect of disability glare on contrast is given in appendix 5.

**Fig 1.16** Discomfort glare from bright luminaires

**Fig 1.17** Disability glare from reflections

## 1.7.2 Discomfort glare from electric lighting

The discomfort experienced when some elements of an interior have a much higher luminance than others can be immediate but sometimes may only become evident after prolonged exposure. The degree of discomfort experienced will depend on the luminance and size of the glare source, the luminance of the background against which it is seen and the position of the glare source relative to the line of sight. A high source luminance, large source area, low background luminance and a position close to the line of sight all increase discomfort glare. Unfortunately most of the variables available to the designer alter more than one factor. For example, changing the luminaire to reduce the source luminance may also reduce the background luminance and these two factors operate in different directions. However, as a general rule, discomfort glare can be avoided by the choice of luminaire and its position, and the use of high reflectance surfaces for the ceiling and upper walls. In any proposed lighting installation the likelihood of discomfort glare being experienced can be estimated by calculating the

Glare Index (see appendix 5). Recommended limiting Glare Indices for specific applications and maximum luminances for luminous ceilings and indirect lighting installations are given in Part 2: Recommendations.

### 1.7.3 Discomfort glare from windows

The most severe visual discomfort which can arise from windows is when the sun is directly visible through them. However, views of a bright unobstructed overcast sky can also be uncomfortable. To overcome these problems it is suggested that the windows be fitted with curtains or blinds which can be used as required. (See section 4.4.1.4.)

# 1.8 Veiling reflections

Veiling reflections are high luminance reflections which overlay the detail of the task (Fig 1.18). Such reflections may be sharp edged or vague in outline, but regardless of form they may affect task performance and cause discomfort. Task performance may be affected because veiling reflections usually reduce the contrast of a task. Discomfort may occur because by reducing the task contrast veiling reflections can make task details difficult to see.

**Fig 1.18**   The effects of veiling reflections

Two conditions have to be met before veiling reflections occur. The first is that part of the task, either task detail or background or both, has to be glossy to some degree. The second is that the part of the interior which is specularly reflected towards the observer, and which is called the offending zone, has to have a higher luminance than other parts of the interior. The most common sources of such high luminances are windows and luminaires. The most generally applicable methods of avoiding veiling reflections are to use matt materials in task areas or to arrange the geometry of the viewing situation so that the luminance of the offending zone is low.

The magnitude of the veiling reflections occurring on a two-dimensional task at a particular position can be quantified by the contrast rendering factor. Details of this quantity are given in appendix 6.

It should be noted that although veiling reflections are usually detrimental to task performance there are some circumstances in which they are useful. Appendix 2 contains examples of the use of high luminance reflections in inspection lighting.

*Veiling reflections*

## 1.9 Flicker

All electric lamps operating from an a.c. supply have an inherent oscillation in light output. The source of the oscillation is different for different lamp types, depending on the physical mechanisms by which the light is produced. For incandescent lamps, the oscillation is usually small because of the thermal inertia of the filament. For discharge lamps the oscillation can be more marked and depends on asymmetry and instability in the arc. For most light sources in most situations, the oscillation is not visible but when it does become visible the effect is called flicker. Flicker is a source of distraction and discomfort to people, particularly as it is easily detected by peripheral vision and thus cannot be readily avoided. Although sensitivity to flicker varies widely between individuals, the main factors which influence its perception are the frequency and the amplitude of the modulation, and the area over which it occurs. Large amplitude modulations occurring over large areas at low frequencies are the most uncomfortable conditions. Small amplitude oscillations over small areas may pass unnoticed.

## 1.10 Stroboscopic effect

The oscillation in light output from a lamp can produce a stroboscopic effect even when the oscillation is not visible. The stroboscopic effect is an illusion which makes a moving object appear as stationary, or to be moving in a different manner from that in which it is really moving. The strength of any stroboscopic effect depends on the frequency, regularity and amplitude of the oscillation in luminous flux relative to the frequency and regularity of the movement of the object. The most dramatic effects will occur when the frequencies are matched or are multiples or sub-multiples of each other and the amplitude of modulation is large. Wherever a significant stroboscopic effect is possible the lighting should always be designed to minimise such effects for reasons of safety (see section 4.4.3.1).

## 1.11 Epilogue

This part of the Code has summarised the ways in which various aspects of lighting affect the performance of tasks, the appearance of the interior and the comfort of the occupants. More detailed discussion of the topics summarised here is available in the publications listed in the bibliography.

The most important point to be made by the above discussion is that there is more to good lighting than illuminance alone. For any installation, the designer should also consider the effects of uniformity of illuminance, the directional distribution of light, the colour and reflectances of room surfaces and objects within the room, the light source colour properties, the extent of disability and discomfort glare and the probability of veiling reflections, flicker and stroboscopic effects occurring. To assess each of these aspects of lighting calls for a careful analysis of the activities carried out in the room and an interpretation of the impression to be conveyed. Thought will also have to be given to the combination of the electric lighting with the daylight available and to integrating the lighting with other building services. If all these factors are considered, then lighting installations which are both suitable and satisfactory for their purpose are likely to result.

## 2.1 Content

The recommendations are presented in three sections. The first consists of a schedule which defines the lighting conditions appropriate for specific interiors/activities. The second consists of complementary and supplementary lighting recommendations which are widely applicable. The third consists of energy targets appropriate to different lighting application areas.

This information is provided for interiors where the lighting is used to aid task performance and/or to create an appropriate and comfortable appearance. Interiors where the main purpose of the lighting is decorative are not considered.

## 2.2 Status

The recommendations are representative of good lighting practice. They are judged to hold an appropriate balance between the benefits of lighting, in terms of task performance and visual comfort, and the costs of lighting, in terms of both equipment and energy costs. They have no statutory standing in their own right but may be adopted by appropriate authorities.

## 2.3 Lighting recommendations for specific interiors/activities

**2.3.1 Introduction**

These recommendations provide:
(a) quantitative guidance on the service illuminances and limiting glare indices appropriate for a wide range of interiors/activities, (b) qualitative advice on the aspects of the interior/activity which should influence the selection of lighting equipment and the design of the lighting installation; (c) notification of statutory and advisory documents relevant to each interior/activity (see section 2.3.2(m)).

The recommendations are contained in a schedule, classified by application. If a specific interior/activity cannot be found in the schedule then appropriate recommendations can be obtained either by using an analogous interior/activity or by using the generic details given in table 2.1 to identify an equivalent interior/activity.

The standard service illuminance given in the schedule for each application should always be converted to a design service illuminance by means of the flow chart in table 2.2. The flow chart allows for some common departures from standard situations to be taken into account.

**2.3.2 Interpretation**

When considering the recommendations in the schedule and the flow chart the following points should be borne in mind:

(a) The service illuminance is the mean illuminance throughout the maintenance cycle of the lighting installation, averaged over the relevant area, which may be a part or the whole of an interior or simply the area of each visual task and its immediate surround, depending on the lighting approach adopted.

(b) The standard service illuminance recommended assumes the task and/or interior is representative of its type. For example, the standard service illuminance recommended for rough sawing in a woodwork shop assumes that the materials being handled, the people doing the work, the duration of the work and the consequences of any mistakes are typical of woodwork shops in general.

*(c)* The design service illuminance results from the application of the modifying factors contained in the flow chart (table 2.2) to the standard service illuminance. The modifying factors allow for departures from the assumed typical circumstances. Specifically, the modifying factors take into account the visual difficulty of the work being done, the duration of the work, the prevalence of visual impairment and the consequences of any mistakes.

*(d)* Where the recommendations clearly refer to an activity rather than an interior in general, the design service illuminance should be provided over a surface appropriate for that activity. This surface may be complex in shape but is usually assumed to be a horizontal, vertical or inclined plane or some combination of them.

*(e)* Where recommendations clearly refer to an activity rather than an interior in general, the design service illuminance need only be provided over the task area and its immediate surroundings. The extent to which it is provided away from the task area will be determined by the type of lighting approach adopted (see section 4.4.2 for a discussion on general, localised and local lighting and section 2.4.3 for recommendations on illuminance ratios for localised and local lighting).

*(f)* Where the recommendations clearly refer to an interior in general rather than an activity, the design service illuminance should be provided over the whole area unless otherwise stated in the notes accompanying the recommendations.

*(g)* Where more than one activity takes place in an interior and a general lighting approach is necessary, then the highest of the design service illuminances recommended for the individual activities should be adopted.

*(h)* Where the design service illuminance is 1500 lx or greater, careful consideration should be given to using local lighting and/or visual aids, e.g. magnifiers.

*(i)* The limiting glare indices specify the degree of discomfort glare which will be acceptable from an overhead lighting installation. They are not applicable to local lighting installations. Any glare indices to be compared with the limiting glare index should be calculated by the methods given in CIBS Technical Memorandum 10 (in preparation).

*(j)* The Glare Index calculated for a lighting installation will vary with viewing position and viewing direction. The recommended limiting glare index for each application should not be exceeded for any viewing position and viewing direction of practical importance.

*(k)* Where more than one activity takes place in an interior, the lowest of the limiting glare indices recommended for the individual activities should be adopted.

(l) In some industrial applications the recommendations refer to hazardous or hostile environments. As used here, a hazardous environment is one where the presence of flammable gas, vapour or dust gives rise to a risk of fire or explosion; a hostile environment is one in which the lighting equipment is subject to mechanical, thermal or chemical attack.

(m) The statutory requirements listed in the schedule are intended to be pertinent but may not be exhaustive. Much of the relevant legislation for a specific area will be contained in more general legislation such as The Health and Safety at Work etc. Act, 1974; the Offices, Shops and Railway Premises Act, 1963; The Factories Act, 1961; The Fire Precautions Act, 1971 and the Protection of Eyes Regulations, 1974, amended 1975.

**Table 2.1    Examples of activities/interiors appropriate for each standard service illuminance**

| Standard Service Illuminance (lx) | Characteristics of the activity/interior | Representative activities/interiors |
|---|---|---|
| 50 | Interiors visited rarely with visual tasks confined to movement and casual seeing without perception of detail. | Cable tunnels, indoor storage tanks, walkways. |
| 100 | Interiors visited occasionally with visual tasks confined to movement and casual seeing calling for only limited perception of detail. | Corridors, changing rooms, bulk stores. |
| 150 | Interiors visited occasionally with visual tasks requiring some perception of detail or involving some risk to people, plant or product. | Loading bays, medical stores, switchrooms. |
| 200 | Continuously occupied interiors, visual tasks not requiring any perception or detail. | Monitoring automatic processes in manufacture, casting concrete, turbine halls. |
| 300 | Continuously occupied interiors, visual tasks moderately easy, i.e. large details >10 min arc and/or high contrast. | Packing goods, rough core making in foundries, rough sawing. |
| 500 | Visual tasks moderately difficult, i.e. details to be seen are of moderate size (5-10 min arc) and may be of low contrast. Also colour judgement may be required. | General offices, engine assembly, painting and spraying. |
| 750 | Visual tasks difficult, i.e. details to be seen are small (3-5 min arc) and of low contrast, also good colour judgements may be required. | Drawing offices, ceramic decoration, meat inspection. |
| 1000 | Visual tasks very difficult, i.e. details to be seen are very small (2-3 min arc) and can be of very low contrast. Also accurate colour judgements may be required. | Electronic component assembly, gauge and tool rooms, retouching paintwork. |
| 1500 | Visual tasks extremely difficult, i.e. details to be seen extremely small (1-2 min arc) and of low contrast. Visual aids may be of advantage. | Inspection of graphic reproduction, hand tailoring, fine die sinking. |
| 2000 | Visual tasks exceptionally difficult, i.e. details to be seen exceptionally small (<1 min arc) with very low contrasts. Visual aids will be of advantage. | Assembly of minute mechanisms, finished fabric inspection. |

# Table 2.2 Flow chart for obtaining the design service illuminance from the standard service illuminance

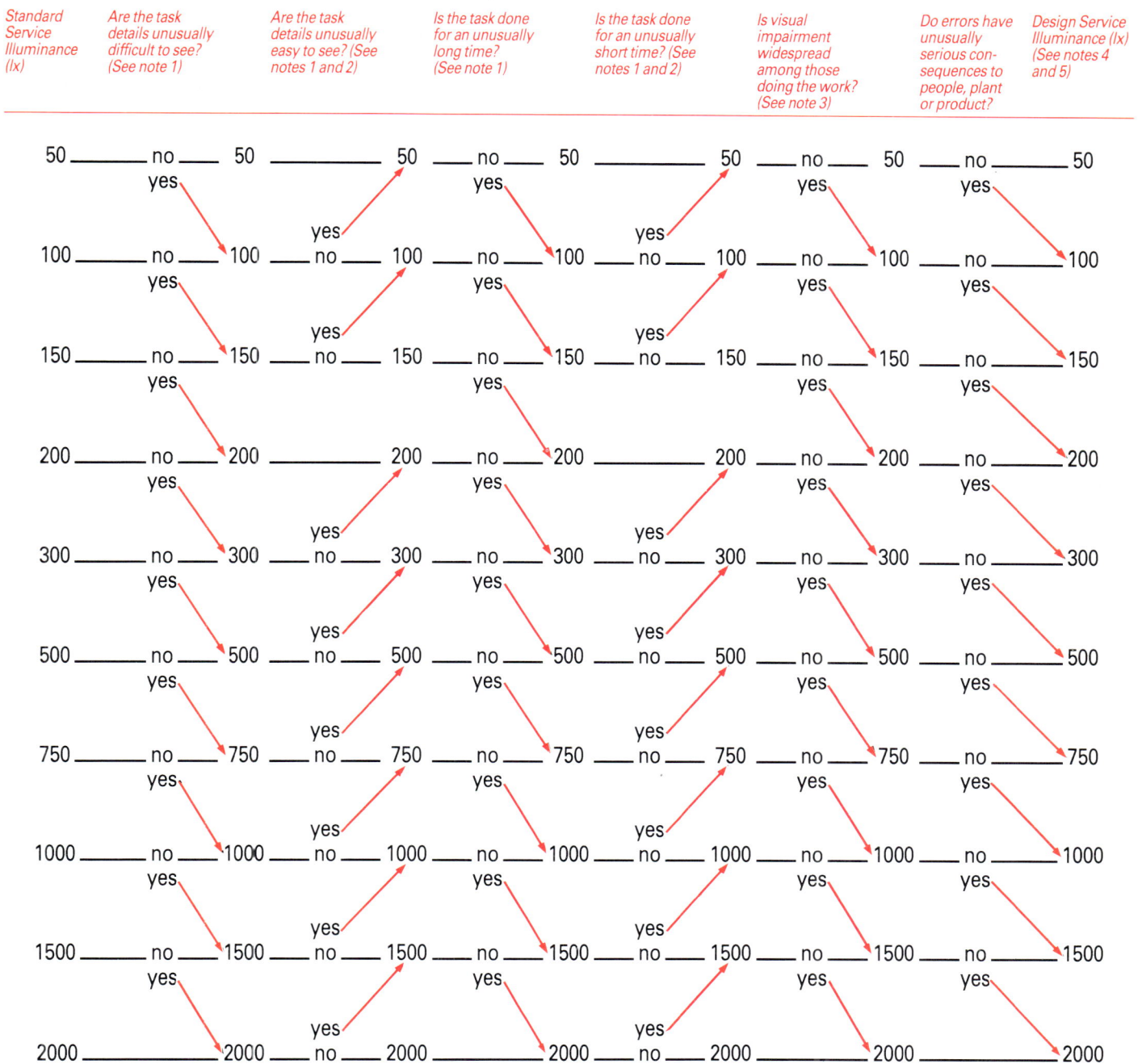

| Standard Service Illuminance (lx) | Are the task details unusually difficult to see? (See note 1) | Are the task details unusually easy to see? (See notes 1 and 2) | Is the task done for an unusually long time? (See note 1) | Is the task done for an unusually short time? (See notes 1 and 2) | Is visual impairment widespread among those doing the work? (See note 3) | Do errors have unusually serious consequences to people, plant or product? | Design Service Illuminance (lx) (See notes 4 and 5) |
|---|---|---|---|---|---|---|---|
| 50 | no → 50 | → 50 | no → 50 | → 50 | no → 50 | no → 50 | 50 |
| 100 | no → 100 | no → 100 | no → 100 | no → 100 | no → 100 | no → 100 | 100 |
| 150 | no → 150 | no → 150 | no → 150 | no → 150 | no → 150 | no → 150 | 150 |
| 200 | no → 200 | → 200 | no → 200 | → 200 | no → 200 | no → 200 | 200 |
| 300 | no → 300 | no → 300 | no → 300 | no → 300 | no → 300 | no → 300 | 300 |
| 500 | no → 500 | no → 500 | no → 500 | no → 500 | no → 500 | no → 500 | 500 |
| 750 | no → 750 | no → 750 | no → 750 | no → 750 | no → 750 | no → 750 | 750 |
| 1000 | no → 1000 | no → 1000 | no → 1000 | no → 1000 | no → 1000 | no → 1000 | 1000 |
| 1500 | no → 1500 | no → 1500 | no → 1500 | no → 1500 | no → 1500 | no → 1500 | 1500 |
| 2000 | → 2000 | no → 2000 | → 2000 | no → 2000 | → 2000 | → 2000 | 2000 |

*(In the flow chart, a "yes" answer to each question moves down one step on the illuminance scale except for the "unusually easy to see" and "unusually short time" columns, where a "yes" answer moves up one step on the illuminance scale.)*

## Notes

1. The standard service illuminances recommended in the schedule are based on tasks which are representative of their type in the detail that has to be seen and the time for which the task has to be done. These steps in the flow chart allow for departures from these assumed conditions.

2. The standard service illuminance of 200 lx is provided as an amenity for continuously occupied interiors, even when perception of task detail is not required.

3. If the cause of visual impairment is dirty or scratched spectacles, safety glasses, safety screens, etc., it may be more effective to clean or replace these items rather than change the lighting. If safety screens are acting as a source of veiling reflections then the lighting/task/worker geometry should be re-arranged.

4. If the design service illuminance is more than two steps on the illuminance scale above the standard service illuminance, consideration should be given to whether changes in the task details, the organisation of the work or the people doing the work are more appropriate than changing the lighting.

5. For a design service illuminance of 1500 lx or 2000 lx, local lighting supplemented by optical aids should be considered.

**2.3.3 Explanatory Notes**

The following notes provide background information on the recommendations in the lighting schedule and flow chart:

(a) The service illuminances have been selected from the following Commission Internationale de l'Eclairage (CIE) scale: 20, 30, 50, 75, 100, 150, 200, 300, 500, 750, 1000, 1500, 2000 lx. The steps in this scale are spaced so as to give a noticeable difference in the subjective effect produced by a uniform lighting installation.

(b) The standard service illuminances are based on considerations of the performance of appropriate tasks, the comfort of the people doing the tasks, and the time for which the space is occupied. Where tasks involve the perception of small detail the service illuminances are determined by the need to ensure quick, accurate and easy task performance. When the tasks require only limited perception of detail, the service illuminances are determined by the duration of the occupation of the space. For continuously occupied spaces, where no perception of detail is required a service illuminance of 200 lx is recommended as an amenity. For a rarely visited location, where no perception of detail is required an illuminance of 50 lx is recommended. If the occupation of the space is intermittent and limited perception of detail is required 100 lx is appropriate. Where perception of detail is necessary but moderately easy, then a service illuminance of 300 lx is recommended for spaces which are continually occupied and 150 lx for intermittently occupied spaces.

(c) The flow chart allows for departures from the assumed circumstances for four different aspects of a specific situation. These are: (i) the visual demands of the task, (ii) the duration of the work; (iii) the visual capabilities of the people doing the work; (iv) the consequences of any errors.

The standard service illuminance given in the schedule assumes the task is representative of its type. If the task is much more visually difficult than usual, e.g. smaller size, lower contrast, then an increase in the service illuminance is appropriate. Conversely, if the task detail is such that the task is easier to see than usual, e.g. larger size, higher contrast, a reduction in service illuminance can be made.

The standard service illuminance given in the schedule assumes the task is to be done over a conventional working period. If the work is to be done continually for a much longer period than usual, the service illuminance should be increased in order to diminish the risk of visual fatigue. Conversely, if the work is to be done over a much shorter period than usual, the service illuminance may be reduced provided the likely increase in time taken to do the task is acceptable.

The standard service illuminance given in the schedule assumes that the people doing the work have normal visual capabilities. If a significant number of people using the lighting have some degree of visual impairment, the service illuminance should be increased. Visual impairment may arise from the use of dirty and scratched spectacles, safety lenses or safety screens, or simply from the deterioration of the visual system with age. Dirty and scratched spectacles, safety glasses and safety screens reduce the transmission of light to the eye and increase the scattered component of what light is transmitted. The most common effects of old age on vision are an increase in the shortest distance at which an object can be focussed, a reduced light transmission through the eye and an increase in the scattering of light in the eye. Spectacles or contact lenses can be used to correct the first effect. Increasing the service illuminance will offset the loss in transmission and will increase the sensitivity of the visual system. Eliminating sources of glare will reduce the effects of light scattered in the eye.

Although increasing the service illuminance and avoiding glare will benefit most people with some degree of visual impairment, it should be noted that there are some severe forms of visual defect, e.g. cataract, for which such changes in lighting may be detrimental to visual abilities. In this situation, great care is necessary to match the lighting conditions to the nature of the visual defect.

The standard service illuminance given in the schedule assumes that the consequences of any errors made are typical of the activity. However, if any errors have unusually serious consequences for people, plant or product then an increase in the service illuminance is appropriate.

(d) The recommended limiting glare indices are based on formal assessments of actual installations and on accumulated experience of the degree of discomfort glare acceptable in different situations. The limiting glare indices recommended have been selected from the following scale: 16, 19, 22, 25 and 28. A difference of 3 glare index units is necessary for a practically significant change in discomfort glare sensation to occur.

## 2.3.4  Schedule

# Index

The activities/interiors listed in the schedule are grouped under 22 main headings.

Subsidiary headings are given in the following alphabetical list:

# Agriculture and Horticulture

**Other relevant documents**

1  The Agricultural (Safety Health and Welfare Provisions) Act, 1956 and associated regulations. 2  The Milk and Dairies (General) Regulations, 1959. 3  Technical Report on Farm Lighting, Ministry of Agriculture, Fisheries and Food, 1976. 4  BS 5502 Design of Buildings and Structures for Agriculture. 5  Farm Electric 25, Lighting for Farms, Electricity Council. 6  Grow electric 1, Growing rooms, Electricity Council. 7  Grow electric 2, Lighting in greenhouses, Electricity Council. 8  CIBS Application Guide, Lighting in Hostile and Hazardous Environments.

**Equipment requirements**

Luminaires should be positioned where mechanical damage is unlikely and be capable of withstanding dirty conditions. In some farm buildings an explosive dust hazard may exist; in these circumstances special equipment may be required. In some areas, luminaires capable of being safely hosed down are needed (IP X5). Low voltage switching is desirable.

**Inspection of farm produce**   The choice of light source is important (see appendix 2).

| | Standard Service Illuminance (lx) | Limiting Glare Index | Notes |
| --- | --- | --- | --- |
| Where colour is important | 500 | — | Local lighting may be appropriate |
| Other inspection tasks | 300 | — | Local lighting may be appropriate |

**Farm workshops**   Eye protection will be required for some activities

| | | | |
| --- | --- | --- | --- |
| General | 100 | 25 | Supplementary local lighting will be necessary |
| Workbench or machine | 300 | — | Local or portable lighting may be appropriate |

| | | | |
| --- | --- | --- | --- |
| **Milk premises** | 100 | 25 | Luminaires suitable for being hosed down may be required in some areas |
| **Sick animal pens, calf nurseries** | 50 | — | A lower illuminance is acceptable in the absence of the stockman |

**Other farm and horticultural buildings**

| | | | |
| --- | --- | --- | --- |
| Where adequate daylight is admitted | 20 | 25 | A 5% average daylight factor will give adequate daylight |
| All other buildings | 50 | 25 | For a windowless building a standard service illuminance of 20 lx may be used provided the building is entered through a vestibule which excludes daylight and which has a standard service illuminance of 50 lx |

# Fuel Industries

## Coal Mining (Surface buildings)

**Other relevant documents**

1  Mines and Quarries Act 1954 and associated Regulations. 2  Mining Industry Act 1920 and associated Regulations. 3  National Coal Board Underground Lighting Recommendations. 4  National Coal Board Design Guide for Coal Preparation Plants. 5  CIBS Application Guide: Lighting in Hostile and Hazardous Environments.

**Equipment requirements**

Luminaires suitable for hazardous atmospheres will be required in some areas. Dirt and damp are likely to produce difficult maintenance conditions in some areas. Luminaires with good maintenance properties are desirable as is easy access to them.

**Coal preparation plant**   Severe fouling of luminaires is likely. Luminaires suitable for being hosed down may be required

| | Standard Service Illuminance (lx) | Limiting Glare Index | Notes |
|---|---|---|---|
| Walkways, floors under conveyors | 50 | — | |
| Wagon loading, bunkers | 50 | — | |
| Elevators, chute transfer pits, washbox area | 100 | — | |
| Drum filters, screens, rotating shafts | 150 | — | |
| Picking belts | 200 | — | Directional and colour properties of the lighting may be important for easy recognition of coal and rock |
| **Lamp rooms** | | | |
| Repair section | 300 | 25 | Supplementary local lighting may be necessary |
| Other areas | 150 | 28 | |
| **Weigh cabins, fan houses** | 150 | 28 | |
| **Winding houses** | 150 | 25 | |

# Gas Manufacture, Storage and Distribution

**Other relevant documents**

1  The Chemical Works Regulations (1922). 2  British Gas Engineering Standard PS/EL1. 3  CIBS Application Guide: Lighting in Hostile and Hazardous Environments. 4  CIBS Technical Memorandum 6: Lighting and visual display units.

**Equipment requirements**

Luminaires certified as suitable for hazardous areas will be needed where explosive gas-air mixtures may arise.

**Coke ovens**   Severe fouling of luminaires is likely in some areas

| | Standard Service Illuminance (lx) | Limiting Glare Index | Notes |
|---|---|---|---|
| Oven top | 30 | — | |
| Gas alleys | 50 | — | |
| Platforms and walkways | 50 | — | |
| Quenching station | 50 | — | |
| Paddle feeder pit | 100 | — | |
| **Control rooms** | 300 | 19 | Localised lighting of the control display and the control desks may be appropriate. In all cases, care should be taken to avoid shadows and veiling reflections occurring on instruments. Care is also required to avoid reflections from visual display units. Where a mainly self luminous display is used the ability to dim the room lighting may be useful. Where a large mimic diagram containing detail which has to be seen from a considerable distance is used, special lighting providing a minimum illuminance of 500 lx should be provided on the appropriate surface of the control panels. For maintenance purposes an illuminance of at least 150 lx should be provided on the appropriate surface of the control panels. |

| | Standard Service Illuminance (lx) | Limiting Glare Index | Notes |
|---|---|---|---|
| Relay and telecommunication rooms | 300 | 25 | |
| Switchrooms | 150 | — | Additional local lighting of instruments and controls may be required |
| Pump houses, water treatment plant house | 150 | — | |
| Standby generator rooms, compressor rooms | 150 | — | Additional local lighting of instruments and controls may be required |
| Off-take/pressure reduction stations | | | |
| Indoor | 150 | — | Additional local lighting of instruments and controls may be required |
| Outdoor | 30 | — | |
| Storage tanks (indoor) | 50 | — | |
| and operating areas and filling points at outdoor tanks | | | |

# Electricity Generation, Transmission and Distribution

**Other relevant documents**
1 The Electricity Regulations, 1908. 2 Central Electricity Generating Board, Standard 12701, CEGB, 1981. 3 CIBS Technical Memorandum 6: Lighting and visual display units. 4 CIBS Application Guide: Lighting in hostile and hazardous environments

**Equipment requirements**
Robust luminaires are needed to withstand vibration in some areas. A damp and dirty environment requiring appropriate luminaires will exist in some parts of the plant.

General Plant

| | Standard Service Illuminance (lx) | Limiting Glare Index | Notes |
|---|---|---|---|
| Turbine houses (operating floor) | 200 | 25 | ⎰ Additional local lighting of instruments may be ⎱ required |
| Boiler and turbine house basements | 100 | 25 | |
| Boiler houses, platforms, areas around burners | 100 | — | |
| Switchrooms, meter rooms, oil plant rooms H.V. substations (indoor) | 150 | — | Additional local lighting of instruments and controls may be required |
| Control rooms | 300 | 19 | Localised lighting of the control display and the control desks may be appropriate. In all cases, care should be taken to avoid shadows and veiling reflections occurring on instruments. Care is also required to avoid reflections on VDUs. Where a mainly self-luminous display is used the ability to dim the room lighting may be useful. Where a large mimic diagram containing detail which has to be seen from a considerable distance is used, special lighting providing a minimum illuminance of 500 lx on the diagram is desirable. For maintenance |

| | Standard Service Illuminance (lx) | Limiting Glare Index | Notes |
|---|---|---|---|
| | | | purposes, an illuminance of at least 150 lx should be provided on the appropriate surface of the control panels. |
| Relay and telecommunication rooms | 300 | 25 | |
| Diesel generator rooms, compressor rooms | 150 | — | Additional local lighting of instruments and controls may be required |
| Pump houses, water treatment plant houses | 150 | — | |
| Battery rooms, chargers, rectifiers | 100 | — | Corrosive atmosphere possible |
| Precipitator chambers, platforms, etc. | 100 | — | |
| Cable tunnels and basements, circulating water culverts and screen chambers, storage tanks (indoors), operating areas and filling points at outdoor tanks | 50 | — | |
| **Coal plant** | | | |
| Conveyors, gantries, junction towers, unloading hoppers, ash-handling plants, settling pits, dust hopper outlets | 100 | — | Severe fouling of luminaires is likely |
| Other areas where operators may be in attendance | 150 | — | |
| **Nuclear plant** | | | |
| Gas circulation bays, reactor areas, boiler platform, reactor charge and discharge faces | 150 | 25 | |

# Metal Manufacture

**Other relevant documents**
1   The Protection of Eyes Regulations (1974) (Amended 1975). 2   British Steel Corporation, Corporate Engineering Standard CES 35, 1981. 3   CIBS Application Guide: Lighting in Hostile and Hazardous Environments

**Equipment requirements**
Lamps and luminaires may be subject to severe fouling and vibration and a wide range of ambient temperatures. Corrosive conditions may be present in some locations. Dustproof (IP5X) or other luminaires with good maintenance properties are desirable, as is easy access to them. The accurate visual judgement of hot metal may be difficult under high pressure sodium discharge lamps. For some metals, dusts formed during production may represent an explosion hazard. Luminaires should be chosen appropriately. Eye protection will be required in some areas.

### Ironmaking
| | Standard Service Illuminance (lx) | Limiting Glare Index | Notes |
|---|---|---|---|
| Sinter Plant: | | | |
|   Plant floor | 200 | 28 | Supplementary lighting may be needed for maintenance work |
|   Mixer drum, fan house, screen houses, coolers, transfer stations | 150 | 28 | |
| Furnaces, cupolas: | | | |
|   General | 150 | 28 | |
|   Control platforms | 300 | — | Local lighting may be appropriate |
| Conveyor galleries, walkways | 50 | 28 | |

*Lighting recommendations for specific interiors/activities*

| | Standard Service Illuminance (lx) | Limiting Glare Index | Notes |
|---|---|---|---|
| **Steelmaking** | | | |
| Electric melting shops | 200 | 28 | ⎰ Supplementary lighting may be needed for |
| Basic oxygen steelmaking plants: | | | ⎱ maintenance work |
|     General | 150 | 28 | |
|     Convertor floor, teeming bay | 200 | 28 | |
|     Control platforms | 300 | — | Local lighting may be appropriate |
| Scrap bays | 150 | 28 | |
| **Metal forming and treatment** | | | |
| Ingot stripping, soaking pits, annealing and heat treatment bays, acid recovery plant | 200 | 28 | Supplementary lighting may be needed for maintenance work. Luminaires capable of withstanding corrosive atmospheres may be needed |
| Pickling and cleaning bays, roughing mills, cold mills, finishing mills, tinning and galvanising lines, cut up and rewind lines: | | | |
|     General | 150 | 28 | |
|     Control platforms | 300 | — | Local lighting may be appropriate |
| Wire mills, product finishing, steel inspection and treatment | 300 | 28 | |
| Plate/strip inspection | 500 | 25 | |
| Inspection of tinplate, stainless steel, etc. | — | — | Special lighting to reveal faults in the specular surface of the material will be required (see appendix 2) |
| **Foundries** | | | |
| Automatic plant: | | | |
|     without manual operation | 50 | 28 | ⎰ Supplementary lighting may be needed for |
|     with occasional manual operation | 150 | 28 | ⎱ maintenance work |
|     with continuous manual operation | 200 | 28 | |
| Control room | 300 | 19 | Localised lighting of the control display and the control desks may be appropriate. In all cases, care should be taken to avoid shadows and veiling reflections occurring on instruments. Care is also required to avoid reflections on VDU's. Where a mainly self-luminous display is used the ability to dim the room lighting may be useful. Where a large mimic diagram containing detail which has to be seen from a considerable distance is used, special lighting providing a minimum illuminance of 500 lx on the diagram is desirable. For maintenance purposes, an illuminance of at least 150 lx should be provided on the appropriate surface of the control panels. |
| Control platforms | 300 | — | Local lighting may be appropriate |
| Non-automatic plant: | | | |
|     charging floor, pouring, shaking out, cleaning, grinding, fettling | 300 | 28 | If blast cleaning is used the luminaires should be away from the work area. Where metal castings are cleaned by means of abrasive wheels or bands, the dust produced may represent an explosion hazard; luminaires should be chosen appropriately. |

| | Standard Service Illuminance (lx) | Limiting Glare Index | Notes |
|---|---|---|---|
| Rough moulding, rough core making | 300 | 28 | Light distribution needs to be diffused and flexible to ensure good lighting of deep moulds. If coloured moulding sands are used, high pressure sodium discharge lamps may not be suitable |
| Fine moulding, fine core making | 500 | 25 | |
| Inspection | 500 | — | See appendix 2 |
| **Forges** Severe vibration is likely to occur | | | |
| General | 300 | 25 | |
| Inspection | 500 | — | See appendix 2 |

# Ceramics

**Other relevant documents**

1   The Protection of Eyes Regulations (1974) (amended 1975). 2   The Clay Works (Welfare) Special Regulations, 1948

**Equipment requirements**

Lamps and luminaires may be subject to severe fouling and high ambient temperatures in some areas. Dustproof (IP5X) or other luminaires with good maintenance properties are desirable in areas where raw materials are formed into the basic product.

| | Standard Service Illuminance (lx) | Limiting Glare Index | Notes |
|---|---|---|---|
| **Concrete products**  Eye protection will be required for some activities | | | |
| Mixing, casting, cleaning | 200 | 28 | |
| **Potteries**  Eye protection will be required for some activities | | | |
| Grinding, moulding, pressing, cleaning, trimming, glazing, firing glazing, firing | 300 | 28 | |
| Enamelling, colouring | 750 | 16 | Where good colour judgements are necessary lamps of colour rendering groups 1A or 1B are desirable |
| **Glass works**  Eye protection will be required for some activities | | | |
| Furnace rooms, bending, annealing lehrs | 150 | 28 | |
| Mixing rooms, forming, cutting, grinding, polishing, toughening | 300 | 28 | Supplementary local lighting may be appropriate. Care should be taken to avoid specular reflections in the work pieces |
| Bevelling, decorative cutting, etching, silvering | 500 | 22 | Supplementary local lighting is necessary. Indirect background lighting is desirable to avoid specular reflections in the work pieces |
| Inspection | 500 | — | See appendix 2 |

# Chemicals

**Other relevant documents**

1   The Chemical Works Regulations, 1922. 2   The Patent Fuel Manufacture (Health & Welfare) Special Regulations 1946. 3   The Highly Flammable Liquids and Liquified Petroleum Gases Regulations, 1972. 4   The Protection of Eyes Regulations 1974 (amended 1975). 5   CIBS Application Guide: Lighting in Hostile and Hazardous Environments.

| | Standard Service Illuminance (lx) | Limiting Glare Index | Notes |
|---|---|---|---|

## Equipment requirements

Many activities in the chemical industry involve a risk of corrosion, fire or explosion. In areas where these risks exist, luminaires appropriate to the nature of the hazard are required. Steam and/or dust may be continually present in some areas and frequent maintenance may be required. Luminaires should be chosen appropriately. Where accurate colour judgements are needed, lamps of colour rendering groups 1A or 1B should be used unless otherwise stated.

### Petroleum, chemical and petrochemical works

Eye protection will be required for some activities

| | Standard Service Illuminance (lx) | Limiting Glare Index | Notes |
|---|---|---|---|
| Exterior walkways, platforms, stairs and ladders | 50 | — | |
| Exterior pump and valve areas | 100 | — | |
| Pump and compressor houses | 150 | — | |
| Process plant with remote control | 50 | — | Supplementary local lighting may be needed for maintenance work |
| Process plant requiring occasional manual intervention | 100 | — | |
| Permanently occupied work stations in process plant | 200 | — | |
| Control rooms for process plant | 300 | 19 | Localised lighting of the control display and the control desks may be appropriate. In all cases, care should be taken to avoid shadows and veiling reflections occurring on instruments. Care is also required to avoid reflections on VDUs. Where a mainly self-luminous display is used the ability to dim the room lighting may be useful. Where a large mimic diagram containing detail which has to be seen from a considerable distance is used, special lighting providing a minimum illuminance of 500 lx on the diagram is desirable. For maintenance purposes, an illuminance of at least 150 lx should be provided on the appropriate surface of the control panels. |

### Pharmaceutical and fine chemicals manufacture

Eye protection will be required for some activities

| | Standard Service Illuminance (lx) | Limiting Glare Index | Notes |
|---|---|---|---|
| Pharmaceutical manufacture: | | | |
| grinding, granulating, mixing, drying, tableting, sterilising, washing, preparation of solutions, filling, capping, wrapping, hardening | 500 | 22 | Clean room technology will often apply |
| Fine chemical manufacture: | | | |
| Exterior walkways, platforms, stairs and ladders | 50 | — | |
| Process plant | 100 | 25 | Supplementary local lighting may be needed for maintenance work |
| Fine chemical finishing | 500 | 25 | |
| Inspection | 500 | — | Local lighting may be applicable. Lamps of colour rendering groups 1A or 1B are desirable (see Appendix 2) |

### Soap manufacture

Luminaires suitable for damp conditions may be required

| | Standard Service Illuminance (lx) | Limiting Glare Index | Notes |
|---|---|---|---|
| General area | 300 | 25 | |
| Automatic processes | 200 | 25 | |

| | Standard Service Illuminance (lx) | Limiting Glare Index | Notes |
|---|---|---|---|
| Control panels | 300 | — | Local lighting may be appropriate. Care should be taken to avoid veiling reflections from instrument displays |
| Machines | 300 | 25 | |
| **Paint works** | | | |
| General | 300 | 25 | If colour judgement is used on the production line, lamps of colour rendering groups 1A or 1B are desirable |
| Automatic processes | 200 | 25 | |
| Control panels | 300 | — | Local lighting may be appropriate. Care should be taken to avoid veiling reflections from instrument displays |
| Special batch mixing | 750 | 22 | ⎰ For exact colour matching work lamps of colour |
| Colour matching | 1000 | 19 | ⎱ rendering group 1A are necessary. BS950 applies |

# Mechanical Engineering

**Other relevant documents**

1 The Protection of Eyes Regulations 1974 (amended 1975). 2 The Horizontal Milling Machines Regulations 1928. 3 The Grinding of Cutlery and Edge Tools Regulations 1925. 4 The Grinding of Metals (Miscellaneous Industries) Regulations 1925. 5 The Highly Flammable Liquids and Liquified Petroleum Gases Regulations 1972. 6 CIBS Application Guide: Lighting in Hostile and Hazardous Environments. 7 Protection against Ultraviolet Radiation in the Workplace, National Radiological Protection Board.

**Equipment requirements**

Robust, easily maintained luminaires are desirable. For many operations local lighting is desirable to provide directional effects on the workpiece. Where brass, copper or similarly coloured materials are used, care should be taken to ensure that high pressure sodium discharge lamps enable the task details to be adequately discriminated. Some areas may represent corrosion or fire/explosion hazards. Luminaires appropriate to these conditions are necessary.

**Structural steel fabrication**   Eye protection will be required for some activities

| | | | |
|---|---|---|---|
| General | 300 | 28 | |
| Marking off | 500 | 28 | Local lighting may be appropriate |

**Sheet metal works**   Eye protection will be required for some activities

| | | | |
|---|---|---|---|
| Pressing, punching, shearing, stamping, spinning, folding | 500 | 22 | |
| Benchwork, scribing, inspection | 750 | 22 | Care should be taken to avoid multiple shadows. Where scribing coatings are used, care should be taken to ensure that high pressure sodium discharge lamps allow the task to be adequately discriminated |

**Machine and tool shops**   Eye protection will be required for some activities

| | | | |
|---|---|---|---|
| Rough bench and machine work | 300 | 25 | ⎰ Some obstruction is likely. Care should be taken to |
| Medium bench and machine work | 500 | 22 | ⎟ minimise stroboscopic effects on rotating |
| Fine bench and machine work | 750 | 22 | ⎟ machinery. Supplementary local lighting on machines is desirable |
| Gauge rooms | 1000 | 19 | Optical aids may be required |

**Die sinking shops**   Eye protection will be required for some activities

| | | | |
|---|---|---|---|
| General | 500 | 22 | |
| Fine work | 1500 | — | Flexible local lighting is desirable |

| | Standard Service Illuminance (lx) | Limiting Glare Index | Notes |
|---|---|---|---|

### Welding and soldering shops   Eye protection will be required

| | Standard Service Illuminance (lx) | Limiting Glare Index | Notes |
|---|---|---|---|
| Gas and arc welding, rough spot welding | 300 | 28 | Care is necessary to prevent exposure of eyes and skin to radiation |
| Medium soldering, brazing, spot welding | 500 | 25 | Welding screens will be used so considerable obstruction is likely. Portable lighting may be useful. |
| Fine soldering, fine spot welding | 1000 | — | Local lighting is desirable |

### Assembly shops   Eye protection will be required for some activities

| | Standard Service Illuminance (lx) | Limiting Glare Index | Notes |
|---|---|---|---|
| Rough work, e.g. frame and heavy machinery assembly | 300 | 25 | Considerable obstruction likely, portable local lighting may be useful. The lighting of vertical surfaces may be important |
| Medium work, e.g. engine assembly, vehicle body assembly | 500 | 22 | Some obstruction likely |
| Fine work, e.g. office machinery assembly | 750 | 19 | Local or localised lighting may be appropriate |
| Very fine work, e.g. instrument assembly | 1000 | — | Local lighting and optical aids are desirable |
| Minute work, e.g. watch making | 1500 | — | Local lighting and optical aids are desirable |

### Inspection and testing shops

| | Standard Service Illuminance (lx) | Limiting Glare Index | Notes |
|---|---|---|---|
| Coarse work, e.g. using go/no go gauges, inspection of large sub-assemblies | 500 | 22 | Local or localised lighting may be appropriate. Use of lamps of colour rendering groups 1A or 1B is recommended if colour judgements are important |
| Medium work, e.g. inspection of painted surfaces | 750 | 19 | |
| Fine work, e.g. using calibrated scales, inspection of precision mechanisms | 1000 | 19 | |
| Very fine work e.g. gauging and inspection of small intricate parts | 1500 | — | Local lighting and optical aids are desirable. Use of lamps of colour rendering groups 1A or 1B is recommended if colour judgements are important |
| Minute work e.g. inspection of very small instruments | 2000 | — | |

### Paint shops and spray booths

The Highly Flammable Liquids and Petroleum Gases Regulations (1972) and/or Local Authority regulations may apply. Luminaires suitable for a hazardous environment are normally necessary. Eye protection will be required for some activities. Frequent maintenance will be required if luminaires are likely to be sprayed during painting operations

| | Standard Service Illuminance (lx) | Limiting Glare Index | Notes |
|---|---|---|---|
| Dipping, rough spraying | 300 | 25 | |
| Preparation, ordinary painting, spraying and finishing | 500 | 22 | Lamps of colour rendering groups 1A or 1B are desirable |
| Fine painting, spraying and finishing | 750 | 22 | |
| Inspection, retouching, matching | 1000 | 22 | Lamps of colour rendering group 1A are desirable (see appendix 2) |

### Plating shops   Eye protection will be required for some activities. Easily maintained luminaires suitable for a humid, corrosive atmosphere are desirable

| | Standard Service Illuminance (lx) | Limiting Glare Index | Notes |
|---|---|---|---|
| Vats and baths | 300 | 25 | |
| Buffing, polishing, burnishing | 500 | 22 | |
| Final buffing and polishing | 750 | 22 | |
| Inspection | — | — | See Appendix 2 |

| | Standard Service Illuminance (lx) | Limiting Glare Index | Notes |
|---|---|---|---|

# Electrical and Electronic Engineering

**Other relevant documents**
1  The Electricity Regulations 1908. 2  The Wireless Telegraphy (Control of interference from fluorescent lighting apparatus) Regulations 1978. 3  The Protection of Eyes Regulations 1974 (amended 1975).

**Equipment requirements**
Where colour judgements are important, lamps of colour rendering groups 1A or 1B are desirable. Where copper, brass and similarly coloured materials are used, care should be taken to ensure that high pressure sodium discharge lamps enable task details to be adequately discriminated.

| Electrical equipment manufacture | Standard Service Illuminance (lx) | Limiting Glare Index | Notes |
|---|---|---|---|
| Eye protection will be required for some activities | | | |
| Manufacture of cables, and insulated wires, winding, varnishing and immersion of coils, assembly of large machines, simple assembly work | 300 | 25 | For large machines, some obstruction is likely, portable local lighting may be needed. |
| Medium assembly, e.g. telephones, small motors | 500 | 25 | Local lighting may be appropriate |
| Assembly of precision components, e.g. telecommunication equipment; adjustment, inspection and calibration | 1000 | — | Local lighting is desirable. Care is necessary to control specular reflections. Optical aids may be useful |
| Assembly of high precision parts | 1500 | — | Local lighting is desirable. Care is necessary to control specular reflections. Optical aids may be useful |

| Electronic equipment manufacture | Standard Service Illuminance (lx) | Limiting Glare Index | Notes |
|---|---|---|---|
| Eye protection will be required for some activities | | | |
| Printed circuit boards: | | | |
|   Silk screening | 500 | — | Local Lighting may be appropriate |
|   Hand insertion of components, soldering | 750 | — | Local Lighting may be appropriate |
|   Inspection | 1000 | — | A large, low luminance luminaire overhead ensures specular reflection conditions which are helpful for inspection of printed circuits (see appendix 2). |
| Assembly of wiring harness, cleating harness, testing and calibration | 750 | — | Local lighting may be appropriate |
| Chassis assembly | 1000 | — | Local lighting may be appropriate |
| Inspection and testing: | | | |
|   Soak test | 200 | 25 | |
|   Safety and functional tests | 300 | 25 | Care should be taken to avoid veiling reflections from instrument displays |

# Food, Drink and Tobacco

**Other relevant documents**
1  The Food Hygiene (General) Regulations, 1970. 2  The Herring Curing Welfare Order, 1927 (Scotland 1926). 3  The Slaughterhouse (Hygiene) Regulations, 1958. 4  The Milk and Dairies (General) Regulations, 1959. 5  CIBS Application Guide: Lighting in Hostile and Hazardous Environments.

| | Standard Service Illuminance (lx) | Limiting Glare Index | Notes |
|---|---|---|---|

## Equipment requirements

Equipment requirements vary widely with the application. The Food Hygiene (General) Regulations require that there be no likelihood of any part of the luminaire or lamp falling into foodstuffs. Therefore in some locations any openable part of a luminaire should be hinged or connected with chains and all fixing should be captive. Further, the lamps should be enclosed to ensure that accidental lamp breakage does not allow debris to fall into the product. In some areas luminaires capable of being washed or hosed down (IPX5) are desirable. Some locations involve fire or explosion risks. For some applications lamps and circuitry which will operate efficiently at high or low temperatures and in humid atmospheres are necessary. Other areas will be wet and/or dusty. In all cases the chosen luminaire should be appropriate for the operating conditions.

### Slaughterhouses

Damp conditions may be present, hosing down may be part of the cleaning process. Luminaires should be chosen appropriately

| | Standard Service Illuminance (lx) | Limiting Glare Index | Notes |
|---|---|---|---|
| General | 500 | 25 | Statutory minimum illuminance = 20 lumen/ft$^2$ (215 lx) |
| Inspection | 750 | 19 | Statutory minimum illuminance = 50 lumen/ft$^2$ (540 lx). Lamps of colour rendering group 1A are required |

### Canning, preserving and freezing

Hosing down may be part of the cleaning procedure for some areas

| | Standard Service Illuminance (lx) | Limiting Glare Index | Notes |
|---|---|---|---|
| Grading and sorting of raw materials | 750 | 22 | The choice of light source is important if colour judgement is required (see Appendix 2) |
| Preparation | 500 | 25 | |
| Canned and bottled goods: | | | |
|   Retorts | 300 | 25 | Some obstruction is likely. Warm/humid conditions may be present. Supplementary lighting may be necessary for cleaning and maintenance |
|   Automatic processes | 200 | 25 | |
| Labelling and packaging | 300 | 25 | |
| Frozen foods: | | | |
|   Process area | 300 | 25 | |
|   Packaging and storage | 300 | 25 | For cold stores see Distribution and Storage |

### Bottling, brewing and distilling

Distilleries may contain areas which are a fire and explosion hazard. Warm, humid conditions may be present in some areas. Luminaires should be chosen appropriately

| | Standard Service Illuminance (lx) | Limiting Glare Index | Notes |
|---|---|---|---|
| Keg washing and handling, bottle washing | 200 | 28 | |
| Keg inspection | 300 | 25 | Optical aids may be used for internal inspection |
| Bottle inspection | — | — | See appendix 2 |
| Process areas | 300 | 25 | |
| Bottle filling | 750 | 25 | |

### Edible oils and fats processing

Areas containing a fire or explosion hazard may be present. Luminaires should be chosen appropriately

| | Standard Service Illuminance (lx) | Limiting Glare Index | Notes |
|---|---|---|---|
| Refining and blending | 300 | 25 | |
| Production | 500 | 22 | |

### Mills

Areas containing a dust explosion hazard may be present. Luminaires should be chosen appropriately. Dust deposition on luminaires may cause maintenance problems

| | Standard Service Illuminance (lx) | Limiting Glare Index | Notes |
|---|---|---|---|
| Milling, filtering and packing | 300 | 25 | |

### Bakeries

| | Standard Service Illuminance (lx) | Limiting Glare Index | Notes |
|---|---|---|---|
| General | 300 | 22 | |
| Hand decorating, icing | 500 | 22 | |

| | Standard Service Illuminance (lx) | Limiting Glare Index | Notes |
|---|---|---|---|
| **Chocolate and confectionery manufacture** | | | |
| General | 300 | 25 | |
| Auomatic processes | 200 | 25 | Supplementary lighting may be necessary for cleaning and maintenance |
| Hand decoration, inspection, wrapping and packing | 500 | 22 | If accurate colour judgements are required lamps of colour rendering groups 1A or B should be used |
| **Tobacco processing** | | | |
| Material preparation, making and packing | 500 | 22 | |
| Hand processes | 750 | 22 | |

# Textiles

**Other relevant documents**
1 The Mule Spinning (Health) Special Regulations, 1953. 2 BS 950, Artificial Daylight for the Assessment of Colour. 3 CIBS Application Guide: Lighting in Hostile and Hazardous Environments.

**Equipment requirements**
Where accurate colour judgements are necessary lamps of colour rendering groups 1A or 1B are desirable. Dirty conditions are likely in many areas of fibre production. In these areas dustproof (IP5X) luminaires are desirable.

| | Standard Service Illuminance (lx) | Limiting Glare Index | Notes |
|---|---|---|---|
| **Fibre preparation** | | | |
| Bale breaking, washing | 300 | 25 | Supplementary lighting may be needed for machine maintenance. Luminaires suitable for a corrosive atmosphere may be required |
| Stock dyeing, tinting | 300 | 25 | Use lamps of colour rendering groups 1A or 1B if colour judgements are necessary |
| **Yarn manufacture** | | | |
| Spinning, roving, winding, etc. | 500 | 22 | Lighting should be concentrated on the thread lines |
| Healding (drawing in) | 1000 | — | Use local lighting |
| **Fabric production** | | | |
| Knitting | 500 | 22 | |
| Weaving | | | |
|   Jute and hemp | 300 | 25 | |
|   Heavy woollens | 500 | 19 | |
|   Medium worsteds, fine woollens, cottons | 750 | 19 | |
|   Fine worsteds, fine linen, synthetics | 1000 | 19 | |
|   Mending | 1500 | 16 | |
|   Inspection | 1500 | — | See appendix 2 |
| **Fabric finishing** | | | |
| Dyeing | 300 | 25 | Supplementary local lighting to 500 lx using lamps of colour rendering groups 1A or 1B is desirable in colour matching areas |
| Calendering, chemical treatment, etc. | 500 | 22 | |
| Inspection: | | | |
|   'Grey' cloth | 1000 | — | Use lamps of colour rendering groups 1A or 1B (see appendix 2). BS 950 should be consulted |
|   Final | 2000 | — | |

*Lighting recommendations for specific interiors/activities*

| | Standard Service Illuminance (lx) | Limiting Glare Index | Notes |
|---|---|---|---|
| **Carpet manufacture** | | | |
| Winding, beaming | 300 | 25 | |
| Setting pattern, tufting, cropping, trimming, fringing, latexing and latex drying | 500 | 22 | |
| Designing, weaving, mending | 750 | 22 | |
| Inspection: | | | |
| General | 1000 | 19 | ⎰ Use local lighting with lamps of colour rendering |
| Piece dyeing | 750 | — | ⎱ groups 1A or 1B (see appendix 2) |

# Leather Industry

**Other relevant documents**
1   CIBS Application Guide: Lighting in Hostile and Hazardous Environments.

**Equipment requirements**
A highly corrosive humid atmosphere will exist in some areas of leather manufacture. Luminaires designed to cope with such a hostile environment are necesary.

## Leather manufacture

| | Standard Service Illuminance (lx) | Limiting Glare Index | Notes |
|---|---|---|---|
| Cleaning, tanning and stretching, vats, cutting, fleshing, stuffing | 300 | 25 | |
| Finishing, scarfing | 500 | 25 | |

## Leather working

| | Standard Service Illuminance (lx) | Limiting Glare Index | Notes |
|---|---|---|---|
| General | 300 | 25 | |
| Pressing, glazing | 500 | 22 | |
| Cutting, splitting, scarfing, sewing | 750 | 22 | Directional lighting may be useful |
| Grading, matching | 1000 | — | Use local lighting with lamps of colour rendering group 1A. If glossy leathers are being examined, a large area, low brightness luminaire should be used to minimise veiling reflections (see appendix 2) |

# Clothing and Footwear

**Other relevant documents**
1   CIBS Application Guide: Lighting in Hostile and Hazardous Environments

**Equipment requirements**
Wherever accurate colour judgements are required, lamps of colour rendering groups 1A or 1B should be used.

## Clothing manufacture

| | Standard Service Illuminance (lx) | Limiting Glare Index | Notes |
|---|---|---|---|
| Preparation of cloth | 300 | 22 | Supplementary lighting will be needed for inspecting cloth (see appendix 2) |
| Cutting | 750 | 19 | |
| Matching | 750 | 19 | Use lamps of colour rendering group 1A |
| Sewing | 1000 | 19 | Supplementary local lighting should be provided on the machines |
| Pressing | 500 | 22 | |
| Inspection | 1500 | 16 | Use lamps of colour rendering groups 1A or 1B |
| Hand tailoring | 1500 | — | Local lighting may be appropriate |

| | Standard Service Illuminance (lx) | Limiting Glare Index | Notes |
|---|---|---|---|
| **Hosiery and knitwear manufacture** | | | |
| Flat bed knitting machines | 500 | 22 | |
| Circular knitting machines | 750 | 22 | Additional local lighting may be required |
| Lockstitch and over-locking machines | 1000 | 19 | |
| Linking or running on | 1000 | 19 | |
| Mending, hand finishing | 1500 | — | Use local lighting |
| Inspection | 1500 | — | Use local lighting with lamps of colour rendering groups 1A or 1B |
| | | | |
| **Glove manufacture** | | | |
| Sorting and grading | 750 | 19 | Use lamps of colour rendering groups 1A or 1B |
| Pressing, knitting, cutting | 500 | 22 | |
| Sewing | 750 | 22 | Supplementary local lighting should be provided on the machines |
| Inspection | 1500 | — | Use local lighting with lamps of colour rendering groups 1A or 1B |
| | | | |
| **Hat manufacture** | | | |
| Stiffening, braiding, refining, forming, sizing, pouncing, ironing | 300 | 22 | |
| Cleaning, flanging, finishing | 500 | 22 | |
| Sewing | 750 | 22 | Supplementary local lighting should be provided on the machines |
| Inspection | 1500 | — | Use local lighting with lamps of colour rendering groups 1A or 1B |
| | | | |
| **Boot and shoe manufacture** | | | |
| Leather and Synthetics: | | | |
|   Sorting and grading | 1000 | 16 | Use lamps of colour rendering groups 1A or 1B. Directional lighting may be useful |
|   Clicking, closing | 1000 | 22 | |
|   Preparatory operations | 1000 | 22 | Local or localised lighting may be appropriate |
|   Cutting tables and presses | 1500 | 16 | |
|   Bottom stock preparation, lasting, bottoming, finishing, shoe rooms | 1000 | 19 | |
| Rubber: | | | |
|   Washing, compounding, coating, drying, varnishing, vulcanising, calendering, cutting | 300 | 25 | Drying may involve an explosion risk. Appropriate luminaires are necessary |
|   Lining, making and finishing | 500 | 22 | |

# Timber and Furniture

**Other relevant documents**

1  The Protection of Eyes Regulations 1974 (amended 1975). 2  The Woodworking Machines Regulations, 1974. 3  The Highly Flammable Liquids and Liquified Petroleum Gases Regulations, 1972. 4  CIBS Application Guide: Lighting in Hostile and Hazardous Environments.

| | Standard Service Illuminance (lx) | Limiting Glare Index | Notes |
|---|---|---|---|

**Equipment requirements**

Dusty conditions are likely anywhere where timber is machined. For these areas dust-tight (IP6X) luminaires are desirable. Luminaires should be cleaned regularly. Wherever rotating machinery is used care should be taken to minimise any stroboscopic effects produced by the lighting.

| | Standard Service Illuminance (lx) | Limiting Glare Index | Notes |
|---|---|---|---|
| **Sawmills** | Eye protection will be required in some areas. Care should be taken to minimise any stroboscopic effects | | |
| General | 200 | 25 | |
| Head saw | 500 | — | Local lighting may be appropriate |
| Grading | 750 | — | Directional lighting may be useful |
| **Woodwork shops** | Eye protection will be required in some areas. Care should be taken to minimise any stroboscopic effects | | |
| Rough sawing, bench work | 300 | 22 | |
| Sizing, planing, sanding, medium machining and benchwork | 500 | 22 | Dust from sanding may represent an explosion hazard; luminaires should be chosen appropriately |
| Fine bench and machine work, fine sanding, finishing | 750 | 22 | Localised lighting may be appropriate. Dust from sanding may represent an explosion hazard; luminaires should be chosen appropriately |
| **Furniture manufacture** | Eye protection will be required in some areas | | |
| Raw materials stores | 100 | 28 | |
| Finished goods stores | 150 | 25 | |
| Wood matching and assembly, rough sawing, cutting | 300 | 22 | Care should be taken to minimise any stroboscopic effects |
| Machining, sanding and assembly, polishing | 500 | 22 | Care should be taken to minimise any stroboscopic effects. The materials used in polishing may represent a fire hazard, and dust from sanding may represent an explosion hazard. Appropriate luminaires will be required |
| Tool rooms | 500 | 22 | |
| Spray booths: | | | The Highly Flammable Liquids and Liquified Petroleum Gases Regulations (1972) and/or Local Authority Regulations may apply. Luminaires should be chosen appropriately |
| Colour finishing | 500 | — | ⎰ Eye protection will be required for some activities. Lamps of colour rendering groups 1A or 1B are desirable |
| Clear finishing | 300 | — | ⎱ |
| Cabinet making: | | | |
| Veneer sorting and grading | 1000 | 19 | Use lamps of colour rendering groups 1A or 1B. Directional lighting may be useful |
| Marquetry, pressing, patching and fitting | 500 | 22 | |
| Final inspection | 750 | — | See appendix 2 |
| **Upholstery manufacture** | | | |
| Cloth inspection | 1500 | — | Use local lighting with lamps of colour rendering groups 1A or 1B (see Appendix 2) |
| Filling, covering | 500 | 22 | |
| Slipping, cutting, sewing | 750 | 22 | |
| Mattress making: | | | |
| Assembly | 500 | 22 | |
| Tape edging | 1000 | 22 | Local lighting may be appropriate |

# Paper and Printing

**Other relevant documents**

1 Lighting in Printing Works, British Printing Industries Federation 1980. 2 CIBS Technical Memorandum 6, Lighting for visual display units. 3 BS 950. artifical daylight for the assessment of colour. 4 CIBS Application Guide: Lighting in Hostile and Hazardous Environments. 5 Protection against ultraviolet radiation in the workplace, National Radiological Protection Board, 1977.

**Equipment requirements**

Damp and dirty conditions are likely in paper mills. Easily maintained, corrosion resistant luminaires are desirable for this application. If volatile inks are used in printing, a hazardous environment may be present in some areas. Luminaires should be chosen appropriately.

## Paper mills

| | Standard Service Illuminance (lx) | Limiting Glare Index | Notes |
| --- | --- | --- | --- |
| Pulp mills, preparation plants | 300 | 25 | |
| Paper and board making: | | | |
|   General | 300 | 25 | |
|   Automatic processes | 200 | 25 | Supplementary lighting may be necessary for maintenance work |
|   Inspection, sorting | 500 | 19 | |
| Paper converting processes: | | | |
|   General | 300 | 25 | |
|   Associated printing | 500 | 22 | When ultraviolet radiation is used for curing ink, special care is necessary to avoid exposure of eyes and skin |

## Printing works

| | Standard Service Illuminance (lx) | Limiting Glare Index | Notes |
| --- | --- | --- | --- |
| Type foundries: | | | |
|   Matrix making, dressing type, hand and machine casting | 300 | 25 | |
|   Font assembly, sorting | 750 | 22 | |
| Composing rooms: | | | Large area, low luminance luminaires are desirable |
|   Hand composing, imposition and distribution | 750 | 19 | |
|   Hot metal – keyboard | 750 | 19 | |
|   Hot metal – casting | 300 | 22 | |
|   Photocomposing – keyboard or setters | 500 | 19 | See CIBS TM6: Lighting for visual display units |
|   Paste up | 750 | 16 | |
| Illuminated tables – general lighting | 300 | — | Use lamps of colour rendering groups 1A or 1B where colour judgements are important, dimming may be required |
| Proof presses | 500 | 22 | Where ultraviolet radiation is used for curing ink, special care is necessary to prevent exposure of eyes and skin |
| Proof reading | 750 | 16 | |
| Graphic reproduction: | | | |
|   General | 500 | 22 | |
|   Precision proofing, retouching, etching | 1000 | — | Local lighting may be appropriate. BS 950 should be consulted where colour is important |
|   Colour reproduction and inspection | 1500 | — | |

| | Standard Service Illuminance (lx) | Limiting Glare Index | Notes |
|---|---|---|---|
| **Printing machine room:** | | | |
| Presses | 500 | 22 | When ultraviolet radiation is used for curing ink, special care is necessary to prevent exposure of eyes and skin |
| Pre-make ready | 500 | 22 | |
| Printed sheet inspection | 1000 | 19 | |
| **Binding:** | | | |
| Folding, pasting, punching, stitching | 500 | 22 | |
| Cutting, assembling, embossing | 750 | 22 | |

# Plastics and Rubber

## Other relevant documents

1 The Chemical Works Regulations (1922). 2 The Protection of Eyes Regulations 1974 (amended 1975). 3 CIBS Application Guide: Lighting in Hostile and Hazardous Environments. 4 Protection against ultraviolet radiation in the workplace, National Radiological Protection Board, 1977.

## Equipment requirements

Some areas in plastic and rubber production may represent a fire, explosion, or corrosion hazard. Luminaires appropriate to these conditions are necessary. Dirty conditions are likely in rubber processing factories. Dustproof (IP5X) or other luminaires with good maintenance properties are desirable for such applications. For both rubber and plastic high ambient temperatures may occur in some areas. Lamps and circuitry capable of operating in high temperatures are desirable in such areas.

**Plastic products**   Eye protection will be required for some activities

| | Standard Service Illuminance (lx) | Limiting Glare Index | Notes |
|---|---|---|---|
| **Automatic plant:** | | | |
| Without manual control | 50 | 28 | Supplementary lighting may be needed for maintenance work |
| With occasional manual control | 100 | 28 | |
| With continuous manual control | 300 | 28 | |
| Control rooms | 300 | 19 | Localised lighting of the control display and the control desks may be appropriate. In all cases, care should be taken to avoid shadows and veiling reflections occurring on instruments. Where a mainly self-luminous display is used the ability to dim the room lighting may be useful. Where a large mimic diagram containing detail which has to be seen from a considerable distance is used, special lighting providing a minimum illuminance of 500 lx on the diagram is desirable. For maintenance at least 150 lx should be provided on the appropriate surface of the control panels |
| Control platforms | 300 | — | Local lighting may be appropriate |
| **Non-automatic plant:** | | | |
| Mixing, calendering, extrusion, injection, compression and blow moulding, sheet fabrication | 300 | 25 | |
| Trimming, cutting, polishing, cementing | 500 | 22 | |
| Printing, inspection | 1000 | 19 | Use lamps of colour rendering groups 1A or 1B when colour judgements are important. When ultraviolet radiation is used to cure inks or lacquers, care is necessary to prevent exposure of eyes and skin |

| | Standard Service Illuminance (lx) | Limiting Glare Index | Notes |
|---|---|---|---|
| **Rubber products** | | | |
| Stock preparation – plasticising, milling | 200 | 25 | |
| Calendering, fabric preparation, stock cutting | 500 | 25 | |
| Extruding, moulding, curing | 500 | 22 | |
| Inspection | 1000 | — | Use local lighting |

# Distribution and Storage

**Other relevant documents**
1   The Chemical Works Regulations (1922). 2   CIBS Application Guide: Lighting in Hostile and Hazardous Environments.

**Equipment requirements**
In some stores, the materials being stored, e.g., chemicals, gases, etc., may produce a corrosive atmosphere or represent a fire/explosion hazard. Luminaires appropriate to these conditions will be required. In cold stores, special luminaires and circuitry may be necessary for some light sources to operate at low temperatures.

**Work stores**   Eye protection will be required for some activities

| | Standard Service Illuminance (lx) | Limiting Glare Index | Notes |
|---|---|---|---|
| Loading bays | 150 | — | Avoid glare to drivers of vehicles approaching the loading bay. Care should be taken to light and mark the edge of the loading bay clearly |
| Unpacking, sorting | 200 | 25 | |
| Large item storage | 100 | 25 | Supplementary local lighting may be necessary if identification of items requires perception of detail. Selective switching may be appropriate |
| Small item rack storage | 300 | 25 | Supplementary local lighting may be necessary if the identification of items is visually difficult. Considerable obstruction is likely. Selective switching may be appropriate |
| Issue counter, records, storeman's desks | 500 | 22 | Local or localised lighting may be appropriate. Care should be taken to avoid reflections on display screens |

**Warehouses and bulk stores**   Eye protection will be required for some activities

| | Standard Service Illuminance (lx) | Limiting Glare Index | Notes |
|---|---|---|---|
| Storage of goods where identification requires only limited perception of detail | 100 | 25 | Lighting should be designed to emphasise the features which enable the operator to identify the required item and its position. The lighting of vertical surfaces will be important. Considerable obstruction is likely. If the storage area is continuously occupied, the standard service illuminance should be increased to 200 lx and to 300 lx where identification requires perception of detail. Avoid glare to forklift truck operators |
| Storage of goods where identification requires perception of detail | 150 | 25 | |
| Automatic high bay rack stores: | | | |
|   Gangway | 20 | — | Supplementary lighting may be required for maintenance |
|   Control station | 200 | — | Avoid glare to operator |
| Packing and despatch | 300 | 25 | |
| Loading bays | 150 | — | Avoid glare to drivers of vehicles approaching the loading bay. Care should be taken to light and mark the edge of the loading bay clearly |

| | Standard Service Illuminance (lx) | Limiting Glare Index | Notes |
|---|---|---|---|

### Cold stores
Eye protection will be required for some activities. Cold and wet conditions are likely to be present

| | | | |
|---|---|---|---|
| General | 300 | 25 | Care should be taken with the lighting of the entrance and exit areas to avoid a sudden change in illuminance, by day or night. A light source with a warm colour appearance may be preferred |
| Breakdown, make-up and despatch | 300 | 25 | |
| Loading bays | 150 | — | Avoid glare to drivers of vehicles approaching the loading bay. Care should be taken to light and mark the edge of the loading bay clearly |

# Commerce

**Other relevant documents**

1   The Offices, Shops and Railway Premises Act, 1963 and associated regulations. 2   The Guide to the 1963 Offices, Shops and Railway Premises Act, HMSO, 1981. 3   CIBS Technical Memorandum 6, Lighting for visual display units.

**Equipment requirements**

Where air conditioning or mechanical ventilation is required, air-handling luminaires may be appropriate.

### Offices

| | | | |
|---|---|---|---|
| General Offices | 500 | 19 | Special considerations apply when visual display units are to be widely used (see CIBS TM 6). Local lighting may be appropriate |
| Deep plan general offices | 750 | 19 | Special considerations apply when visual display units are to be widely used (See CIBS TM 6). If the light distribution of the installation is arranged so as to effectively light vertical surfaces, a lower illuminance can be satisfactory. Local lighting may be appropriate |
| Computer work stations | 300-500 | 19 | See CIBS TM 6 |
| Conference rooms, executive offices | 500 | 19 | Dimming or switching to permit use of visual aids may be necessary |
| Computer and data preparation rooms | 500 | 19 | See CIBS TM 6 |
| Filing rooms | 300 | 19 | Vertical surfaces may be especially important |

### Drawing offices

| | | | |
|---|---|---|---|
| General | 500 | 16 | |
| Drawing boards | 750 | 16 | Local lighting may be appropriate |
| Computer aided design and drafting | — | — | Special lighting is required |
| Print rooms | 300 | 19 | |

### Banks and building societies
The lighting should be designed to create an appropriate atmosphere

| | | | |
|---|---|---|---|
| Counter, office area | 500 | 19 | See CIBS TM 6 |
| Public area | 300 | 19 | |

# Services

**Other relevant documents**

1 The Protection of Eyes Regulations 1974 (amended 1975). 2 The Petroleum (Consolidated) Act 1928. 3 The Highly Flammable Liquids and Liquified Petroleum Gases Regulations, 1972. 4 CIBS Application Guide: Lighting in Hostile and Hazardous Environments.

**Equipment requirements**

The nature of the equipment used varies widely with the application. For garages, easily maintained luminaires capable of operating where a fire or explosion hazard may exist are necessary in some areas. For laundries and sewage treatment works, luminaires capable of withstanding a corrosive damp atmosphere are necessary.

## Garages
Hazardous area lighting will be required in many areas, e.g. servicing pits, spray booths, pump service areas. The Highly Flammable Liquids and Liquified Petroleum Gases Regulations (1972) and/or Local Authority by-laws may apply. Eye protection is required for some activities

| | Standard Service Illuminance (lx) | Limiting Glare Index | Notes |
| --- | --- | --- | --- |
| Interior parking areas | 30 | 22 | |
| General repair, servicing, washing, polishing | 300 | 22 | Luminaires in pits should be easily cleaned and suitable for hazardous areas. Supplementary local lighting will be necessary |
| Workbench | 500 | 19 | Localised or local lighting may be appropriate |
| Spray booths | 500 | 19 | Lamps of colour rendering groups 1A or 1B are desirable. Luminaires appropriate to a hazardous area are necessary. Portable local lighting may be useful |
| External apron: | | | |
| General | 50 | — | ⎰Luminaires suitable for a hazardous area are |
| Pump area (retail sales) | 300 | — | ⎱appropriate. Care should be taken to avoid glare to |
| Pump area (in house sales) | 150 | — | drivers and neighbouring residents |
| Showrooms | — | — | See Retailing |

## Appliance servicing

| | Standard Service Illuminance (lx) | Limiting Glare Index | Notes |
| --- | --- | --- | --- |
| Workshop: | | | |
| General | 300 | 25 | Supplementary portable local lighting is desirable |
| Workbench | 500 | — | Localised lighting may be appropriate |
| Counter | 300 | — | Localised lighting may be apppropriate |
| Stores | 300 | — | Obstruction is likely, selective switching may be appropriate (see storage and distribution) |

## Laundries
Luminaires suitable for a warm, damp atmosphere are necessary

| | Standard Service Illuminance (lx) | Limiting Glare Index | Notes |
| --- | --- | --- | --- |
| Commercial laundries: | | | |
| Receiving, sorting, washing, drying, ironing, despatch, dry cleaning, bulk machine work | 300 | 25 | |
| Hand ironing, pressing, mending, spotting, inspection | 500 | 25 | |
| Launderettes | 300 | 25 | |

## Sewage treatment works
Luminaires capable of withstanding damp, corrosive conditions will be required in some areas

| | Standard Service Illuminance (lx) | Limiting Glare Index | Notes |
| --- | --- | --- | --- |
| Walkways | 50 | — | |
| Process areas | 100 | 28 | |

# Retailing

**Other relevant documents**

1  The Offices, Shops and Railway Premises Act, 1963. 2  The Food Hygiene (General) Regulations, 1970. 3  CIBS Lighting Guide: The outdoor environment. 4 Lighting for retailers, Electricity Council, 1984.

**Equipment requirements**

The types of lighting equipment used will depend greatly on the approach adopted to displaying the merchandise. For some interiors, e.g. a supermarket, simple uniform lighting is appropriate; for others, e.g. a jewellers, localised lighting creating highlights in the merchandise is more suitable. The recommendations given below refer to interiors where uniform lighting is desirable. Where good colour judgement is considered an advantage then lamps of colour rendering group 1B should be used, such lamps will also enhance the colourfulness of an interior.

| | Standard Service Illuminance (lx) | Limiting Glare Index | Notes |
| --- | --- | --- | --- |
| Small shops with counters | 500 | 19 | The service illuminance should be provided on the horizontal plane of the counter. Where wall displays are used, a similar illuminance on the walls is desirable |
| Small, self-service shops with island displays | 500 | 19 | The service illuminance should be provided on the vertical faces of the display stands |
| Supermarkets, hypermarkets: | | | |
| General | 500 | 22 | The service illuminance should be provided on the vertical faces of the displays |
| Checkout | 500 | 22 | The service illuminance should be provided on the horizontal plane of the conveyor |
| Showrooms for large objects, e.g. cars, furniture | 500 | 19 | The service illuminance should be provided at floor level. For some merchandise, vertical surfaces may also be important |
| Shopping precincts and arcades | 150 | 22 | See CIBS Lighting Guide: The Outdoor Environment |

# Places of Public Assembly

**Other relevant documents**

1  The Theatres Act (1968). 2  The Recommendations for Safety in Cinemas (1955), (amended 1976). 3  BS CP 1007 Maintained lighting for cinemas. 4  Lighting and Wiring of Churches, Church of England Church Information Office, 1981. 5  CIBS Lighting Guide: Libraries. 6  CIBS Lighting Guide: Museums and Art Galleries. 7  CIBS Technical Memorandum: Lighting for visual display units. 8  CIBS Lighting Guide: Sports. 9  The handbook of sport and recreational building design. Sports Council, 1981.

**Equipment requirements**

The type of equipment used will vary widely with application depending on the importance attached to the appearance of the equipment and the desired display effects. For applications where materials sensitive to light are being used, e.g. in libraries, museums and art galleries, the thermal and ultraviolet emission of the lighting installation needs to be considered (see reference 6). For very quiet interiors, e.g. churches, libraries, care is needed to minimise noise emission by the lighting installation. For sports areas, impact resistant luminaires may be desirable. For many applications some form of dimming/switching facility is desirable.

## Assembly rooms

| | Standard Service Illuminance (lx) | Limiting Glare Index | Notes |
| --- | --- | --- | --- |
| Public rooms, village halls, church halls | 300 | 19 | These rooms are often used for many different functions. The lighting should be flexible in the effects it can produce. Selective switching or dimming is desirable |

| | Standard Service Illuminance (lx) | Limiting Glare Index | Notes |
|---|---|---|---|
| **Concert halls, cinemas and theatres** | | | |
| Foyer | 200 | — | |
| Booking office | 300 | — | Local or localised lighting may be appropriate |
| Auditorium | 100 | — | Dimming facilities will be necessary. Special lighting of the aisles is desirable |
| Dressing rooms | 300 | — | Special mirror lighting for make-up may be required |
| Projection room | 150 | — | The service illuminance should be provided on the working side of the projector. The lighting should not detract from the view into the auditorium. Dimming facilities are desirable |
| **Churches** | | | |
| Body of church | 150 | 19 | |
| Pulpit, lectern | 300 | — | Use local lighting |
| Choir stalls | 300 | — | Local lighting may be appropriate |
| Altar, communion table, chancel | 150 | — | Additional lighting to provide emphasis is desirable |
| Vestries | 150 | 22 | |
| Organ | 300 | — | Use local lighting |
| **Hospitals** See CIBS Lighting Guide: Hospital and Health Care Buildings | | | |
| **Hotels** | | | |
| Entrance Halls | 100 | — | The lighting of vertical surfaces is important to the appearance of the space. A mean scalar illuminance of at least 40 lx should be provided |
| Reception, cashiers' and porters' desks | 300 | — | Localised lighting may be appropriate |
| Bars, coffee bars, dining rooms, grill rooms, restaurants, lounges | 50-200 | — | The lighting should be designed to create an appropriate atmosphere. Switching and dimming controls can provide some flexibility in the lighting effects. Local lighting to provide emphasis may be a appropriate in some areas, e.g. cash desks, bar counters. Supplementary lighting may be necessary for cleaning |
| Cloakrooms, baggage rooms | 100 | — | |
| Bedrooms | 50 | — | Supplementary local lighting at the bedhead and near a mirror is desirable |
| Bathrooms | 100 | — | Supplementary local lighting near the mirror is desirable |
| Food preparation and stores, cellars, lifts and corridors | — | — | See 'General Building Areas' |
| **Libraries** See CIBS Lighting Guide: Libraries | | | |
| **Lending library** | | | |
| General | 300 | 19 | |
| Counters | 500 | — | Localised lighting may be appropriate |
| Bookshelves | 150 | — | The service illuminance should be provided on the vertical face at the bottom of the bookstack |
| Reading rooms | 300 | 19 | |
| Reading tables | 300 | 19 | Local lighting may be appropriate. Supplementary lighting or optical aids for the partially sighted should be considered |

| | Standard Service Illuminance (lx) | Limiting Glare Index | Notes |
|---|---|---|---|
| Catalogues: | | | |
| Cards | 150 | — | The service illuminance should be provided on the plane of the cards |
| Microfiche/visual display units | 150 | — | Care should be taken to avoid reflections on the screen (see CIBS TM 6) |

## Reference libraries    Care should be taken to minimise the noise emitted by the lighting installation

| | Standard Service Illuminance (lx) | Limiting Glare Index | Notes |
|---|---|---|---|
| General | 300 | 19 | |
| Counters | 500 | — | Localised lighting may be appropriate |
| Bookshelves | 150 | — | The service illuminance should be provided on a vertical surface at the foot of the bookshelves |
| Study tables, carrels | 500 | 19 | Local lighting may be appropriate, care should be taken to avoid veiling reflections |
| Map room | 300 | 19 | Supplementary local lighting is desirable when using maps |
| Display and exhibition areas: | | | |
| Exhibits insensitive to light | 300 | — | |
| Exhibits sensitive to light, e.g., pictures, prints, rare books in archives | — | — | See CIBS Lighting Guide, Museums and Art Galleries |

## Library workrooms

| | Standard Service Illuminance (lx) | Limiting Glare Index | Notes |
|---|---|---|---|
| Book repair and binding | 500 | 19 | |
| Catalogue and sorting | 500 | 19 | |
| Remote book stores | 150 | — | The service illuminance should be provided on a vertical plane at the foot of the book stack. Switching arrangements should be carefully considered |

## Museums and art galleries    See CIBS Lighting Guide: Museums and Art Galleries

| | Standard Service Illuminance (lx) | Limiting Glare Index | Notes |
|---|---|---|---|
| Exhibits insensitive to light | 300 | — | The lighting will be mainly determined by the display requirements |
| Light sensitive exhibits, e.g. oil and tempera paints, undyed leather, bone, ivory, wood, etc. | 150 | — | This is a maximum illuminance to be provided on the principal plane of the exhibit |
| Extremely light sensitive exhibits, e.g., textiles, water colours, prints and drawings, skins, botanical specimens, etc. | 50 | — | This is a maximum illuminance to be provided on the principal plane of the object. Switching and covering to limit exposure is desirable |
| Conservation studios and workshops | 500 | 19 | Supplementary local lighting is desirable for detailed work. If colour judgement is important, light sources of group 1A should be used. Careful control of illuminance by switching/dimming is desirable |

## Sports facilities    See CIBS Lighting Guide: Sports

| | Standard Service Illuminance (lx) | Limiting Glare Index | Notes |
|---|---|---|---|
| Multi-purpose sports halls | 300-750 | — | The lighting system should be sufficiently flexible to provide lighting suitable for the variety of sports and activities that take place in sports halls. Some of these activities are best lit from the side whilst others require lighting from overhead. For details of the lighting requirements for individual sports see CIBS Lighting Guide: Sports |

# Education

**Other relevant documents**

1 The Education (School Premises) Regulations, 1981. 2 Guidelines for Environmental Design and Fuel Conservation in Educational Buildings. Department of Education and Science, Architects and Building Branch, Design Note 17, 1981. 3 CIBS Lighting Guide: Lecture Theatres. 4 CIBS Lighting Guide: Libraries. 5 CIBS Lighting Guide: Sports.

**Equipment requirements**

Education buildings are usually designed to be lit by daylight whenever and wherever possible. Lighting controls should ensure that the lighting can be easily adjusted to accommodate variation in daylight conditions. Special areas in educational buildings, e.g. in workshops, sports halls, laboratories, etc., need luminaires appropriate to the conditions met in these places.

| | Standard Service Illuminance (lx) | Limiting Glare Index | Notes |
|---|---|---|---|
| **Assembly Halls** | | | |
| General | 300 | 19 | Switching or dimming systems which enable the hall to be used for theatrical or cinematic functions are desirable. If it is proposed to use the hall for examination purposes, the standard service illuminance should be provided as a minimum |
| Platform and stage | — | — | Special lighting to provide emphasis and to facilitate the use of the platform/stage is desirable |
| **Teaching spaces** | | | |
| General | 300 | 19 | DES Design Note 17 which contains statutory requirements under the Education (School Premises) Regulations 1981, specifies (a) a minimum illuminance of 150 lx at any point on the working plane no matter what the light source, (b) a service illuminance of not less than 300 lx where fluorescent lamps are used, (c) where the lighting of a space is achieved by a combination of daylight and electric light a service illuminance of not less than 350 lx will usually be necessary. Also the illuminance on the walls should be from 0.5 to 0.8 of the working plane illuminance. Care should be taken with the lighting of chalk boards to avoid veiling reflections and give uniformity. Facilities for switching and dimming are desirable where visual aids are to be used. Lamps of colour rendering group 1B are desirable |
| **Lecture theatres** See CIBS Lighting Guide: Lecture Theatres | | | |
| General | 300 | 19 | Switching and/or dimming facilities are desirable to allow for the use of visual aids; some light should be provided for the lecturer |
| Demonstration benches | 500 | — | Localised lighting may be appropriate |
| **Seminar rooms** | 500 | 19 | Switching and/or dimming facilities are desirable to allow for the use of visual aids but some lighting should be provided for the lecturer |
| **Art rooms** | 500 | 19 | Lamps of colour rendering groups 1A or 1B should be used. Some form of flexible display lighting is desirable |
| **Needlework rooms** | 500 | 19 | Supplementary local lighting is desirable |

| | Standard Service Illuminance (lx) | Limiting Glare Index | Notes |
|---|---|---|---|
| Laboratories | 500 | 19 | If accurate colour judgements are required lamps of colour rendering groups 1A or 1B should be used. In some laboratories there will be fire and/or chemical hazards. A corrosive atmosphere may also be present. Appropriate luminaires are required. Eye protection will be required for some activities |
| Libraries | 300 | 19 | See CIBS Lighting Guide: Libraries |
| Music rooms | 300 | 19 | Care should be taken to minimise the noise emitted by the lighting system |
| Sports halls | 300 | — | See CIBS Lighting Guide: Sports. Impact resistant luminaires may be required. If the hall is to be used for examination purposes the standard service illuminance should be provided as a minimum |
| Workshops | 300 | 19 | See the appropriate industrial processes in this schedule. Supplementary local lighting may be desirable. Eye protection will be required for some activities |

# Transport

**Other relevant documents**

1   The Offices, Shops and Railways Premises Act, 1963 and associated regulations. 2   Guide to the 1963 Offices, Shops and Railway Premises Act, HMSO 1981. 3   CIBS Lighting Guide: The Outdoor Environment. 4   CIBS Technical Memorandum 6: Lighting and visual display units. 5   Recommended Practice for Lighting of Railway Premises, British Railways Board 1969.

**Equipment requirements**

Some areas in transport facilities are open to the weather. Weatherproof luminaires (IPX4) are desirable for these areas.

## Airports
| | | | |
|---|---|---|---|
| Ticket counters, check-in desks and information desks | 500 | — | Localised lighting may be appropriate. Care should be taken to avoid reflections in visual display unit |
| Departure lounges, other waiting areas | 200 | 19 | The lighting should assist in creating a relaxed atmosphere |
| Baggage reclaim | 200 | 22 | Localised lighting may be appropriate |
| Baggage handling | 100 | 22 | |
| Customs and Immigration Halls | 500 | 22 | Localised lighting may be appropriate |
| Concourse | 200 | 22 | The lighting of vertical surfaces is important to the appearance of the concourse. When general lighting is specified a mean scalar illuminance of at least 80 lx is desirable. Care should be taken with the lighting of flight information boards |

## Railway stations
| | | | |
|---|---|---|---|
| Ticket office | 500 | 19 | Localised lighting over the counter may be appropriate. Care should be taken to avoid reflections in visual display units |
| Information office | 500 | 19 | |
| Parcels office, Left luggage office: | | | |
| General | 100 | 22 | |
| Counter | 200 | 22 | Localised lighting is appropriate |
| Waiting rooms | 200 | 22 | The lighting should assist in creating a relaxed atmosphere |

| | Standard Service Illuminance (lx) | Limiting Glare Index | Notes |
|---|---|---|---|
| Concourse | 200 | 22 | Lighting of vertical surfaces is important to the appearance of the concourse. Where general lighting is specified a mean scalar illuminance of at least 80 lx is desirable. Care should be taken with the lighting of train arrival and departure boards |
| Timetables | 200 | — | Local lighting may be appropriate. |
| Ticket barriers | 200 | — | |
| Platforms (covered) | 50 | — | Care should be taken to light and mark the edge of the platform clearly |
| Platforms (open) | 10 | — | |

## Coach stations
| | | | |
|---|---|---|---|
| Ticket offices | 500 | 19 | Localised lighting over the counter may be appropriate |
| Information offices | 500 | 19 | |
| Left luggage office: | | | |
| General | 100 | 22 | |
| Counter | 200 | 22 | Localised lighting is appropriate |
| Waiting rooms | 200 | 22 | The lighting should assist in creating a relaxed atmosphere |
| Concourse | 200 | 22 | Lighting of vertical surfaces is important to the appearance of the concourse. When a general lighting installation is specified a mean scalar illuminance of at least 80 lx is desirable |
| Timetables | 200 | — | Local lighting is appropriate |
| Loading areas | 150 | — | |

# General Building Areas

## Entrances
| | | | |
|---|---|---|---|
| Entrance halls, lobbies, waiting rooms | 200 | 19 | The lighting of vertical surfaces is important to the appearance of the space. A mean scalar illuminance of 80 lx is recommended. Care should be taken with entrance areas to avoid a sudden change of illuminance between inside and outside by day or night |
| Enquiry desks | 500 | 19 | Localised lighting may be appropriate |
| Gatehouses | 200 | 19 | Flexible switching or dimming facilities and low surface reflectances may be desirable for security reasons |

## Circulation areas
| | | | |
|---|---|---|---|
| Lifts | 100 | — | BS 5655 Part 1 specifies a minimum illuminance of 50 lx on the lift car floor |
| Corridors, passageways, stairs | 100 | 22 | Stairs should be lit to provide a contrast between the treads and the risers. Avoid specular reflections on the treads |
| Escalators, travellators | 150 | — | Escalators should be lit to provide a contrast between the treads and the risers. For both escalators and travellators specular reflections on the treads should be avoided |

| | Standard Service Illuminance (lx) | Limiting Glare Index | Notes |
|---|---|---|---|

## Medical and first aid centres

See CIBS Lighting Guide: Hospital and Health Care Facilities

| | Standard Service Illuminance (lx) | Limiting Glare Index | Notes |
|---|---|---|---|
| Consulting rooms, treatment rooms | 500 | — | Various regulations made under the Factories Act 1961 apply to medical examination rooms |
| Rest rooms | 150 | — | Dimming may be desirable |
| Medical stores | 150 | — | Supplementary lighting may be necessary if perception of fine detail is required |

## Staff rooms

| | | | |
|---|---|---|---|
| Changing, locker and cleaners rooms, cloakrooms, lavatories | 100 | — | Various sections of the Factories Act 1961 and the Offices, Shops and Railway Premises Act 1963 apply to lavatories |
| Rest rooms | 150 | 19 | Lighting should be different in style from the work areas |

## Staff restaurants

| | | | |
|---|---|---|---|
| Canteens, cafeterias, dining rooms, messrooms | 200 | 22 | The lighting should aim to provide a relaxed but interesting atmosphere. Various regulations made under The Factories Act 1961 apply to canteens and messrooms |
| Servery, vegetable preparation, washing up area | 300 | 22 | The Food Hygiene (General) Regulations 1970 apply. Luminaires should be constructed so that no part of the lamp or luminaire can fall into the foodstuffs. The luminaires should be capable of being washed or hosed down in safety. Lamps suitable for operation at low temperatures will be necessary for some food storage areas. Lamps and luminaires suitable for hot humid conditions may be required for some other areas. |
| Food preparation and cooking | 500 | 22 | |
| Food stores, cellars | 150 | 22 | |

## Communications

| | | | |
|---|---|---|---|
| Switchboard rooms | 300 | 19 | Avoid veiling reflections from controls. Too high an illuminance may reduce the visibility of signal lights. Supplementary local lighting may be desirable where directories are used |
| Telephone apparatus room | 150 | 25 | |
| Telex room, post room | 500 | 19 | |
| Reprographic room | 300 | 19 | |

## Building Services

Boiler houses:

| | | | |
|---|---|---|---|
| General | 100 | 25 | |
| Boiler front | 150 | — | |
| Boiler control room | 300 | 19 | Localised lighting of the control display and the control desks may be appropriate. In all cases, care should be taken to avoid shadows and veiling reflections occurring on instruments. Care is also required to avoid reflections on VDUs. Where a mainly self-luminous display is used the ability to dim the room lighting may be useful. Where a large mimic diagram containing detail which has to be seen from a considerable distance is used, special lighting providing a minimum illuminance of 500 lx on the diagram is desirable. For maintenance |

| | Standard Service Illuminance (lx) | Limiting Glare Index | Notes |
|---|---|---|---|
| | | | purposes, an illuminance of at least 150 lx should be provided on the appropriate surface of the control panels |
| Control rooms | 300 | 19 | See note above |
| Mechanical plant room | 150 | 25 | Supplementary portable lighting may be required for maintenance |
| Electrical power supply and distribution rooms | 150 | — | Additional local lighting of instruments and controls may be required |
| Store rooms | 100 | — | |

Car parks    Luminaires suitable for hazardous area lighting may be necessary
Covered car parks:

| | Standard Service Illuminance (lx) | Limiting Glare Index | Notes |
|---|---|---|---|
| Floors | 5-20 | — | Care is required in the positioning of luminaires to avoid glare to drivers and pedestrians |
| Ramps and corners | 50 | — | |
| Entrances and exits | 100 | — | The lighting of exits and entrances should provide a transition zone to avoid sudden changes in illuminance between inside and outside by day or night. Care should be taken to avoid glare to drivers and pedestrians |
| Control booths | 200 | — | Local lighting may be appropriate |
| Outdoor car parks | 5-20 | — | See CIBS Lighting Guide: The outdoor environment |

# 2.4 General Lighting Recommendations

### 2.4.1 Introduction

These recommendations provide supplementary and complementary guidance to that given in the schedule and flow chart (section 2.3). They are applicable to many situations and will ensure visual comfort.

### 2.4.2 Illuminance

The design service illuminance for specific applications can be obtained from the schedule and flow chart in section 2.3. The recommendations given are consistent with the rule that no working space which is to be continuously occupied should have an illuminance of less than 200 lx on the working plane.

### 2.4.3 Illuminance ratios

*(a)* The ratio of the minimum illuminance to the average illuminance over the task area should not be less than 0.8.

*(b)* In an interior with general lighting, the ratio of the average illuminance on the ceiling to the average illuminance on the horizontal working plane should be within the range 0.3 to 0.9 (Fig 2.1).

*(c)* In an interior with general lighting, the ratio of the average illuminance of any wall to the average illuminance on the horizontal working plane should be within the range 0.5 to 0.8 (Fig 2.1).

*(d)* In an interior with localised or local lighting, the ratio of the illuminance on the task area to the illuminance around the task area should not be more than 3:1.

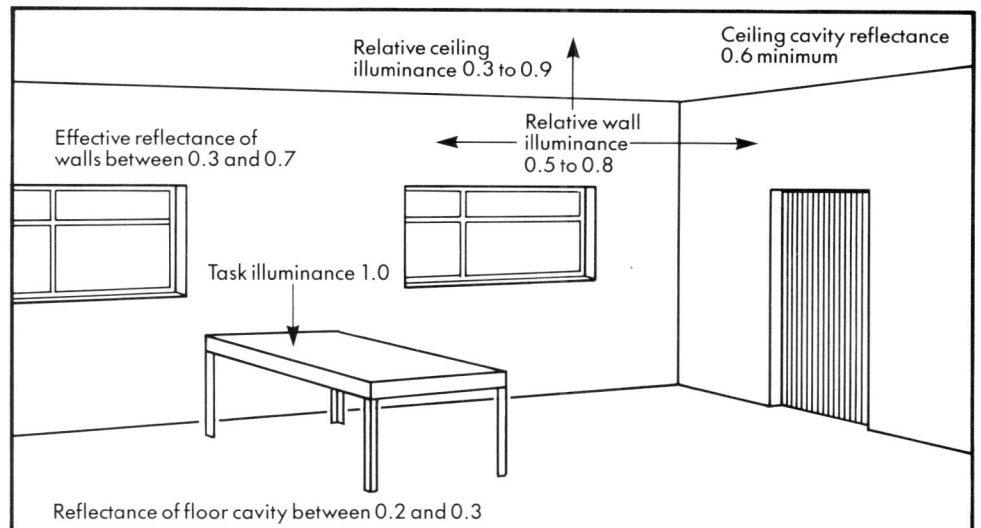

**Fig 2.1** Recommended illuminance ratios and surface reflectances

### 2.4.4 Directional lighting

Directional lighting intended to enhance the appearance of people within a space should have a vector/scalar ratio in the range 1.2-1.8 and should, at least during daytime, have a vector direction of 15-45° below the horizontal (Fig 1.10).

### 2.4.5 Surface reflectances

*(a)* The ceiling cavity reflectance should be as high as practicable and generally at least 0.6. This will usually mean that the reflectance of the paint or other surface finish must be at least 0.8 (Fig 2.1).

(b) The effective reflectance of the principal walls should be between 0.3 and 0.7 (Fig 2.1). This usually means that the wall surface finish will have to have an actual reflectance greater than 0.5. The reflectance of window wall surfaces (including reveals and glazing bars but excluding glazing) should be at least 0.6 to reduce contrast with the bright scene outdoors during daytime. Where the perception of people's faces is particularly important e.g. lecture theatres, conference rooms, the reflectance of walls which form the background against which people are seen should not exceed 0.6.

(c) In general, gloss finishes should not be used over large surface areas. Where indirect lighting is used gloss finishes should not be used anywhere on the ceiling or upper walls.

(d) Where practical, floor cavity reflectance should be within the range 0.2 to 0.3 (Fig 2.1). This usually means that the relevant surfaces will have to have a reflectance greater than 0.3.

(e) It is desirable for equipment and furnishings in working interiors, and in particular desk tops, to have finishes with a reflectance of not less than 0.2.

(f) The immediate background to a task should be matt. In addition, the ratio of the reflectance of the immediate background of a task to that of the task itself should preferably be in the range 0.3 to 0.5.

### 2.4.6 Colour

(a) Where accurate colour judgements are to be made, lamps of colour rendering groups 1A or 1B are necessary (see table 1.1). Where exact colour matching is desired lamps of colour rendering group 1A should be used and the recommendations of BS 950 should be followed, as appropriate. The Munsell Chroma of walls, ceilings and other large surfaces close to colour judgement areas should not be greater than 1 and the reflectances should not be less than 0.4. The illuminance on the task should be at least 500 lx.

(b) Where the objective is to produce a particular effect or to enhance the appearance of a product, the best approach to selecting a suitable light source is direct observation. In general, light sources in colour rendering groups 1A and 1B will make colours appear more colourful than lamps of colour rendering groups 2, 3 or 4.

### 2.4.7. Glare

(a) Luminous ceilings utilising large diffusing panels are not recommended for lighting interiors for which the recommended limiting glare index is less than 19. In any case, the average luminance of such luminous ceiling should not be greater than 500 cd/m$^2$.

(b) For indirect lighting, the average luminance of the ceiling should not be more than 500 cd/m$^2$. However, small areas of luminance up to 1500 cd/m$^2$ will generally be acceptable, provided the edges of the area are not sharply defined.

### 2.4.8 Daylighting

Previous editions of this Code have contained a schedule of recommendations of daylight factor for interiors where daylight from side windows is the chief source of light during the greater part of the day. This Code does not contain such recommendations. The reason for this change is that the previous recommendations are considered to be either too limited or irrelevant. The lighting designer is likely to be faced with the need to consider daylight in two different situations. The first is when the designer is concerned with the daylighting of a new building. The second, and much more common situation, is when the designer is faced with a daylighting installation which is already fixed. In the first case, the recommendations of daylight factor are relevant but insufficient. In the second case daylight factor recommendations are irrelevant. A full treatment of the design aspects of daylighting will be given in a forthcoming CIBS publication. In this Code, the only aspects of daylighting which are considered are those arising from an existing window: namely, how should the electric lighting be related to the natural lighting and does the window need additional visual or thermal protection.

# 2.5 Energy – Installed efficacy recommendations

**2.5.1 Introduction**

These recommendations provide guidance on the energy efficiency of new lighting installations using modern equipment. The recommendations are given in terms of ranges of installed efficacy appropriate for different interiors lit by lamps with different colour properties. The installed efficacy of a lighting installation is defined as:

$$\text{Installed efficacy} = \text{Lamp circuit luminous efficacy} \times \text{Utilisation factor}$$

Installed efficacy is expressed in units of lumens/watt. For tubular fluorescent and other discharge light sources, the lamp circuit luminous efficacy includes the power consumption of both the lamp and its associated control gear. The light output value used in the calculation of lamp circuit luminous efficacy is the lighting design lumen value.

**2.5.2 Status**

The ranges of installed efficacy recommended are representative of good lighting practice. They should be treated as guidelines rather than rigid targets. With careful design it may be possible to achieve a suitable lighting installation at an installed efficacy above the range recommended. Conversely, there are situations where the lighting and architectural requirements justify an installed efficacy below the range recommended.

**2.5.3 Method of use**

The ranges of installed efficacy appropriate for different application areas are given in table 2.3. The ranges are classified according to the application area, the Room Index of the interior, and the Colour Rendering Group of the light source used. The relationship between the actual installed efficacy and the recommended range of installed efficacies should be carefully considered (see 2.5.4 Interpretation). Ideally, the actual installed efficacy will be towards the high end of the recommended range, but achieving this may depend on factors beyond the control of the lighting designer.

**2.5.4 Interpretation**

When considering the recommended ranges of installed efficacy the following points should be borne in mind.

(a) A range of installed efficacies rather than a single value is recommended because the installed efficacy for each specific application will vary with the reflectance of the room surfaces and the cleanliness of the interior. A very dirty interior will have an installed efficacy towards the low end of the range because the reflectances of surfaces within the interior will inevitably be low. Conversely, a very clean interior can have an installed efficacy towards the top end of the range, provided the reflectances of surfaces in the interior are high.

(b) The ranges of installed efficacies apply to general lighting installations, i.e. the same illuminance is provided over the whole working plane. They should not be applied to localised or local lighting systems.

(c) Where special luminaires e.g. flameproof luminaires, are required the range of installed efficacies should be derated by multiplying by 0.7.

(d) Where a Glare Index of less than 19 is required for commercial and retail premises or a Glare Index of less than 22 is required for industrial purposes, the range of installed efficacies should be derated by multiplying by 0.7.

(e) Where considerable obstruction to the lighting is likely to occur, the range of installed efficacies may be derated considerably.

(f) The installed efficacy target can be easily converted to a power loading target, expressed in $W/m^2$ once the illuminance provided is identified. The power loading is given by the expression.

$$\text{Power loading} = \frac{\text{Illuminance}}{\text{Installed efficacy}}$$

where illuminance is expressed in lux and installed efficacy is expressed in lumens/watt

This power loading assumes clean installations, i.e. it has a Maintenance Factor of unity.

## 2.5.5 Explanatory notes

The following notes provide background information on the recommended installed efficacies.

(a) The ranges of installed efficacy have been derived from an examination of current lighting practice using modern lighting equipment.

(b) The classification by application area is used because of the influence of the mounting heights likely to be available and hence the light sources which can be used. It is assumed that high bay industrial buildings allow the use of high power discharge lamps and that other industrial and commercial interiors will have a range of mounting heights which will allow either low power discharge lamps or tubular flourescent lamps to be used. It is also assumed that luminaires suitable to the application will be used.

(c) The classification by room index is used because the proportions of the room influence the proportion of the luminous flux emitted by the lamp which reaches the working plane directly. In rooms with a low room index the working plane receives a large proportion of its luminous flux after reflection from the bounding surfaces. This reduces the utilisation factor and hence decreases the installed efficacy.

(d) Classification by colour rendering group is used because it influences the choice of light source. Four classes are used, based on colour rendering groups 1B, 2, 3 and 4 (see table 1.1). It is assumed that lamps of groups 2, 3 and 4 will be used in high bay industrial premises, lamps of groups 1B, 2, 3 and 4 in other industrial premises and lamps of groups 1B, 2 and 3 only will be used in commercial premises.

(e) Lamps of colour rendering group 1A are only used when very accurate colour matches are essential. This is only likely to occur over small areas. This, together with the low luminous efficacy of these lamps, suggests that lamps of colour rendering group 1A are not likely to be used for general lighting and hence need not be included in table 2.3.

**Table 2.3  Installed efficacy range targets for uniform lighting installations (lm/W)**

| Application Area | Room Index | Lamp Colour Rendering Group | | | |
|---|---|---|---|---|---|
| | | 1B | 2 | 3 | 4 |
| High bay industrial | 1 | | 18-29 | 14-23 | 21-45 |
| | 2 | | 23-37 | 18-29 | 27-55 |
| | 5 | | 27-43 | 20-32 | 32-60 |
| Industrial (not high bay) | 1 | 14-23 | 14-23 | 14-23 | 19-31 |
| | 2 | 18-29 | 18-29 | 18-29 | 23-37 |
| | 5 | 20-32 | 20-32 | 20-32 | 26-42 |
| Commercial | 1 | 14-19 | 14-19 | 14-19 | |
| | 2 | 18-27 | 18-27 | 18-27 | |
| | 5 | 20-30 | 20-30 | 20-30 | |

# Part 3    Equipment

## 3.1 Introduction

The aims of this part of the Code are *(a)* to provide an outline of the properties of the main types of lighting equipment currently used for interior lighting and *(b)* to give details of appropriate maintenance procedures. It must be emphasised that the information given is only sufficient to demonstrate the differences between broad classes of equipment. It is not precise enough for design purposes.

FOR DESIGN WORK, UP-TO-DATE, ACCURATE INFORMATION ON LIGHT SOURCES, LUMINAIRES AND CONTROL SYSTEMS SHOULD BE OBTAINED FROM MANUFACTURERS.

## 3.2 Light sources

### 3.2.1 Types of light source

The main types of light source used are:

*(a)* tungsten filament *(b)* tungsten halogen filament *(c)* high pressure mercury tungsten discharge (blended) *(d)* high pressure mercury discharge (fluorescent) *(e)* high pressure mercury discharge (metal halide) *(f)* low pressure mercury discharge (tubular fluorescent) *(g)* high pressure sodium discharge *(h)* low pressure sodium discharge

Within each type there are a range of lamps available which differ in construction, wattage, luminous efficacy, colour properties, cost, etc. Associated with each discharge lamp type is a set of prefix letters to facilitate identification. Table 3.1 shows the prefix letters adopted in the U.K. for the main types of discharge lamp used in interiors. The prefix letters are usually accompanied by the lamp rating in watts and, if necessary, the details of lamp length, colour and lamp cap type.

### 3.2.2 Lamp characteristics

The construction, operation, range of luminous efficacies, life and colour properties of each lamp type are summarised in table 3.2 along with some typical applications. This table only gives an overview of the range of lamp types available. For information on a specific lamp the manufacturer's data should always be consulted.

#### 3.2.2.1 Construction

Each basic lamp type described in table 3.2 can have a number of variations in construction. These variations can involve its shape, the number and type of caps it has, the presence of a fluorescent or diffusing coating on an outer envelope, the chemical composition of any fluorescent coating and the provision of a reflector inside the lamp. Manufacturer's data should be consulted for details of the options available.

#### 3.2.2.2 Operation

The operating details given in table 3.2 are concerned with such matters as run-up time, re-ignition time, operating positions, and susceptibility to environmental conditions. Run up times and re-ignition times are important because most of the discharge lamps do not produce their maximum light output immediately after switch on. Usually several minutes are required before the maximum light output is achieved. Further, unless special circuits are used, high pressure discharge lamps will not immediately re-ignite after an interruption of supply. Usually a period of several minutes is necessary for the lamp to cool before it will re-ignite. These factors limit the suitability of some lamp types for rapid switching and dimming. It is also worth noting that not all lamp types can operate in all positions and that some lamp types are sensitive to such external environmental factors as air temperature and vibration.

**3.2.2.3 Luminous efficacy**

The luminous efficacies given in table 3.2 are expressed in lumens/lamp watt. The lumen output of each lamp type used in the calculation of luminous efficacy is the lighting design lumen value. Table 3.3 gives a comparison of the ranges of luminous efficacies of the different lamp types.

**3.2.2.4 Lamp life**

The life of an electric lamp can have two distinct meanings. These are *(a)* the time after which the lamp ceases to operate, *(b)* the time after which the light output is so reduced that it is more economic to replace the lamp even though it is still functioning electrically. Typically, lamps with filaments fall into the first category but discharge lamps fall into the second. Whilst defining the average life of lamps with filaments presents little problem,

**Table 3.1    Prefix letters used in the UK to identify types of discharge lamp in common use**

| Prefix Letters | Description of Lamp |
|---|---|
| MBF | A high pressure mercury discharge lamp with an arc tube and a fluorescent coating on the inside of the outer envelope |
| MBFT/MBTF | An MBF lamp with a tungsten filament in series with the arc tube: a 'blended' light source |
| MBFR | An MBF lamp in which part of the outer envelope has an inner reflecting coating |
| MBI | A high pressure mercury discharge lamp with metal halide additives in the arc tube and a clear outer envelope |
| MBIL | An MBI lamp in linear double ended form without an outer envelope |
| MBIF | An MBI lamp with a fluorescent coating on the inside of the outer envelope |
| MCF | A low pressure mercury discharge lamp with a linear glass discharge tube with an internal fluorescent coating: a tubular fluorescent lamp |
| MCFA | An MCF lamp with an external conducting strip connected to both end caps |
| MCFE | An MCF lamp with an external water repellent coating |
| MCFR | An MCF lamp with an internal reflecting coating on part of the tube |
| SON* | A high pressure sodium discharge lamp with an arc tube in an outer envelope |
| SON-TD/SON-L | A linear, double-ended SON lamp with a tubular outer envelope |
| SON-R | A SON lamp with an internal reflecting coating |
| SLI | A low pressure sodium discharge lamp with linear arc tube in which arc tube and outer envelope are combined to form one unit (linear construction) |
| SOX | A low pressure sodium lamp with U shaped arc tube, of integral construction |

*The SON lamp family is developing rapidly. As new types of SON lamp are introduced, new prefix letters will be associated with them. The latest edition of the Lighting Industry Federation Factfinder 3: Lamp Guide, should be consulted for information on the prefix letters used for SON lamps introduced since the publication of this Code.*

defining the life of a discharge lamp does, because it depends so strongly on the economic factors involved. Table 3.2 gives ranges of lamp life for each lamp type. For lamps with filaments the life is expressed as the time after which 50% of a large sample of lamps will have failed. For discharge lamps, with the exception of the low pressure sodium type, life is expressed as the time after which the light output of the lamp will have fallen 30% below the initial light output. For the low pressure sodium discharge lamp, the life is related to a 30% reduction in luminous efficacy, rather than a 30% reduction in light output because for this lamp type the luminous efficacy tends to change with time rather than the light output. A range of times is given for each lamp type because the time for a specific installation will vary with the construction and rating of the lamp used, even for lamps of the same type, and with such operating conditions as the voltage applied and the switching cycle.

### 3.2.2.5 Colour properties

Different lamp types have different colour properties. Further, for some lamp types it is possible to change these colour properties by using fluorescent coatings of different chemical composition. This approach is widely used in the low pressure mercury discharge lamp (tubular fluorescent). The colour properties of each lamp type are characterised by their correlated colour temperature (CCT) class (table 1.1), the colour rendering group (table 1.1) and a brief description based on a visual assessment of how the lamps affect the appearance of colours.

### 3.2.3 Control gear

The control gear which is associated with discharge lamps should fulfill three functions. It should *(a)* start the lamp, *(b)* control the lamp current after ignition, *(c)* correct the power factor. Control gear consumes energy and for a given type, some circuits consume more than others. The efficacy of a lamp circuit as a whole depends on the total power taken by the lamp and the control gear. It is also necessary to consider the power factor of the circuit in order to minimise electricity charges and to ensure correct cable ratings.

The current and wattage ratings of cables, fuses and switchgear used in the control gear must be related to the total current in the circuit, although allowance may be necessary for increased currents and voltages during switching. Harmonic currents may be present and will increase the neutral current in a three-phase system. Current ratings of neutral conductors should be the same as that of phase conductors. Manufacturers can supply information about the power factor and harmonic currents of their control gear. All electrical installations should comply with the current edition, and amendments, of the Regulations for Electrical Installations, published by the Institution of Electrical Engineers.

It is important to appreciate that a lamp and the associated control gear constitute an integrated unit for producing light. Lamps from different manufacturers may not operate on the same control gear even when the lamps are nominally of the same type. Whenever any change is proposed in either element of the lamp/control gear package, care should be taken to ensure that the proposed combination is compatible, both electrically and physically.

It should also be noted that special types of control gear are necessary if dimming or rapid re-ignition of some types of discharge lamp is required.

The life of control gear is sensitive to ambient temperature. The control gear used should have an appropriate temperature rating for the situation. If this temperature is exceeded the insulating material may deteriorate rapidly.

**3.2.4 Innovations and trends**

The information given concerning light sources and control gear is correct at the date of publication. However, rapid developments are taking place in the field of light sources and control gear. For example, it is likely that over the life of this Code, new discharge light sources with high luminous efficacy and good colour properties will be introduced, as will electronic control gear with improved facilities for switch and dimming control. Because of these developments and because of the need to use accurate information in lighting design, the information given here should only be treated as indicative. Definitive information on light sources and control gear should be sought from the manufacturers. A frequently up-dated source of information on lamp types is the Lighting Industry Federation Factfinder 3: Lamp Guide.

**Table 3.2    General characteristics of lamps used for lighting in interiors**

| Lamp type | Construction and Operation | Range of luminous efficacy (lumens/lamp watt) | Life |
|---|---|---|---|
| Tungsten filament | A tungsten filament heated to incandescence in a glass envelope usually filled with an inert gas; does not require any control gear.<br>Immediate full light output, most types operate in all positions, light output and life sensitive to small voltage variations, lamp life sensitive to vibration | typically 8-18 | Limited by failure of the filament, average life 1000-2000 hours according to type |
| Tungsten halogen filament | A tungsten filament heated to incandescence in a small envelope containing halogens; does not require any control gear, but may require low voltages.<br>Immediate full light output, may have restricted operating position, light output and life sensitive to small voltage variations, lamp life sensitive to vibration, the envelope surface is liable to deteriorate if touched with bare hands, almost no decline in light output with time | typically 18-24 | Limited by failure of the filament, average life 2000-4000 hours according to type |
| High pressure mercury tungsten discharge (blended) (MBFT/MBTF) | An electric discharge in a high pressure mercury atmosphere contained in an arc tube in series with a tungsten filament heated to incandescence, the whole contained within a glass envelope with a fluorescent coating. Does not need control gear. Some light output immediately but run-up period to 90% full light output about 4 minutes, re-ignition after about 10 minutes, operating positions restricted, life sensitive to voltage variations and vibration. | typically 10-26 | Limited by failure of the filament, average life 5000-8000 hours |
| High pressure mercury discharge (fluorescent) (MBF, MBFR, MBF de luxe) | An electric discharge in a high pressure mercury atmosphere contained in an arc tube within a glass envelope with a fluorescent coating; needs control gear. Run up period to full light output about 4 minutes. Re-ignition after about 10 minutes unless special circuits used; operates in all positions | typically 36-54* | Likely to be determined by economic factors; life to 30% reduction in light output is typically 5000-10000 hours according to type, rating, switching cycle etc. |
| High pressure mercury discharge (metal halide) (MBI, MBIF, MBIL) | An electric discharge in a high pressure mercury atmosphere with metal halide additives in an arc tube sometimes contained within a glass envelope; the outer envelope may have a fluorescent coating; needs control gear.<br>Run up period to 90% of full light output about 5 minutes; re-ignition after about 10 minutes unless special circuits are used; restricted operating positions | typically 66-84* | Likely to be determined by economic factors; life to 30% reduction in light output typically 5000-10000 hours according to type, rating, switching cycle, etc. |

*When calculating the installed efficacy of an installation using this lamp type, the power consumption of the associated control gear should be included in the estimate of luminous efficacy.

| Colour Properties | | | Typical Applications |
|---|---|---|---|
| CCT Class | Colour Rendering Group | Colour Rendering Characteristics (based on visual assessment) | |
| Warm | 1A | Emphasises reds strongly, yellows and greens to a lesser extent, blues strongly subdued | Homes, hotels, restaurants, for general lighting and for display lighting; emergency lighting |
| Warm | 1A | Emphasises reds strongly, yellows and greens to a lesser extent, blues strongly subdued | Display, area floodlighting, shops |
| Inter-mediate | 3 | Emphasises yellows and blues which shift towards violet, subdues reds | As a replacement for tungsten filment lamps where lamp life is important, e.g. because access is difficult |
| Inter-mediate | 3 | Emphasises yellows and blues which shift towards violet, subdues reds | General lighting in factories, area floodlighting |
| Depending on the choice of chemicals used in the arc tube, the CCT class can be warm, intermediate or cold. Similarly, the colour rendering can vary over a wide range but will usually be better than that of the high pressure mercury discharge (fluorescent) lamp. To establish the colour properties of a specific example of this lamp type it is essential to consult the manufacturers | | | Industrial and commercial lighting, e.g. general lighting in high bay factories, shops and offices; area floodlighting |

**Table 3.2** continued

| Lamp type | Construction and Operation | Range of luminous efficacy (lumens/ lamp watt) | Life |
|---|---|---|---|
| Low pressure mercury discharge (tubular fluorescent) (MCF, MCFA, MCFE, MCFR) | An electric discharge in a low pressure mercury atmosphere contained in a glass tube internally coated with a fluorescent material; needs control gear. There are many sizes of lamps and types of fluorescent coating which produce a wide range of luminous efficacies and colour properties, the most common types are shown in the table. For convenience they be divided into four groups: first a group used for applications where accurate colour judgements are required; second a group suitable for general use and having rare-earth, triphosphor coatings, third a group suitable for general use but having halo-phosphate coatings and fourth a group of compact, low power lamps designed as alternatives to the tungsten filament lamp. Tubular fluorescent lamps are available in a wide range of physical and electrical sizes. This lamp type is the subject of considerable development by the Lighting Industry. The latest edition of the Lighting Industry Federation's Factfinder 3: Lamp guide, should be consulted for information on developments occurring since the publication of this Code.<br><br>Immediate light output and restrike; operates in all positions, light output sensitive to ambient temperature, air temperatures above and below about 25°C reduces the light output; difficult to start at low temperatures.<br><br>All the tubular fluorescent lamps described are of the hot cathode type, i.e. they need a heated cathode to operate. A cold cathode tubular fluorescent lamp also exists. It is widely used for illuminated signs and occasionally for interior lighting. Compared to the hot cathode type, the cold cathode tubular fluorescent lamp has a longer life, a lower luminous efficacy and a higher operating voltage. Its main advantage in use is that it can easily be formed into long, complex shapes. | typically 37-90* | Likely to be determined by economic factors; life to 30% reduction in light output typically 5000-10000 hours according to type, switching cycle, etc. |

*When calculating the installed efficacy of an installation using this lamp type, the power consumption of the associated control gear should be included in the estimate of luminous efficacy

| Lamp Name | CCT Class | Colour Rendering Group | Colour Properties | Typical Applications |
|---|---|---|---|---|
| | | | Colour Rendering Characteristics (based on visual assessment) | |
| **Special lamps** | | | | |
| Northlight/Colour Matching | Cold | 1A | Similar to north skylight; emphasises blues, and to a lesser extent, greens | Used where colour rendering similar to north sky daylight is needed |
| Artificial daylight | Cold | 1A | Similar to Northlight/Colour Matching but emits more ultraviolet to conform with natural skylight | Used where colour matching to BS 950 Part 1 is required |
| Kolor-rite Trucolor 38 | Inter-mediate | 1A | Equal emphasis given to all colours | Used where fine colour judgements are required, e.g. hospitals, art galleries |
| **Tri-phosphor lamps** | | | | |
| 4000K e.g. Colour 84, Energy Saver 84, Polylux 4000 | Inter-mediate | 1B | Emphasises orange, greens and blue-violets, subdues yellows and deep reds | Factories, offices and shops |
| 3000K, e.g. Colour 83, Energy Saver 183, Polylux 3000 | Warm | 1B | Emphasises oranges, greens and blue-violets, subdues yellows and deep reds | Social areas, restaurants, hotels, homes |
| **Halo-phosphate lamps** | | | | |
| Cool White | Inter-mediate | 2 | Emphasises yellow, and to a lesser extent, greens and blues; red shifts slightly towards orange | Factories, offices and shops |
| White | Inter-mediate | 3 | Emphasises yellows, and to a lesser extent, greens; subdues reds and to some extent blues, which shifts towards violet | Factories, offices and shops |
| Warm White | Warm | 3 | Emphasises yellows, and to a lesser extent, greens, reds slightly subdued blues subdued and shifted towards violet | Commercial/public buildings |
| **Low power, compact lamps** | | | | |
| 2D, SL, PL lamps etc. | Warm | 1B | Emphasises oranges, greens, blues and violets, subdues some yellows and deep reds | Hotels, shops, homes; as a replacement for tungsten filament lamps |

**Table 3.2** continued

| Lamp type | Construction and Operation | Range of luminous efficacy (lumens/ lamp watt) | Life |
|---|---|---|---|
| High pressure sodium discharge (SON, SON-TD/SON-L, SON-R) | An electric discharge in a high pressure sodium atmosphere in an arc tube contained in a diffusing or clear outer envelope, needs control gear. Run-up time to 90% of light output from 4 to 7 minutes. Re-ignition within 1 minute if an external ignitor is used. Operates in any position. The high pressure sodium discharge lamp family is developing rapidly. The latest edition of the Lighting Industry Federation Factfinder 3: Lamp Guide, should be consulted for information on developments occurring since the publication of this Code | typically 67-121* | Likely to be determined by economic factors; life to 30% reduction in light output is typically 6000-12000 hours according to rating, switching cycle, etc. |
| Low pressure sodium discharge (SOX, SLI) | An electric discharge in a low pressure sodium atmosphere in a glass arc tube contained in a glass envelope; needs control gear. Run-up to 90% of full light output from 6-12 minutes; re-ignition typically within 3 minutes; restricted operating positions | typically 101-175* | Likely to be determined by economic factors; life to 30% reduction in luminous efficacy is typically 6000-12000 hours depending on lamp construction |

*\* When calculating the installed efficacy of an installation using this lamp type, the power consumption of the associated control gear should be included in the estimate of luminous efficacy*

| CCT Class | Colour Rendering Group | Colour Rendering Characteristics (based on visual assessment) | Typical Applications |
|---|---|---|---|
| Warm | 2 or 4 | Emphasises yellows, reds to a lesser extent, greens are acceptable but blues are subdued | General lighting in factories, warehouses, commercial buildings; area floodlighting |
| | | This is an almost monochromatic light source. It virtually emits only yellow light so all colours except yellow appear brown or black | Industrial area lighting where distortion of most colours is acceptable; road lighting |

*Colour Properties* spans the CCT Class, Colour Rendering Group, and Colour Rendering Characteristics columns. *Typical Applications* is its own heading.

## Table 3.3 Comparison of luminous efficacies

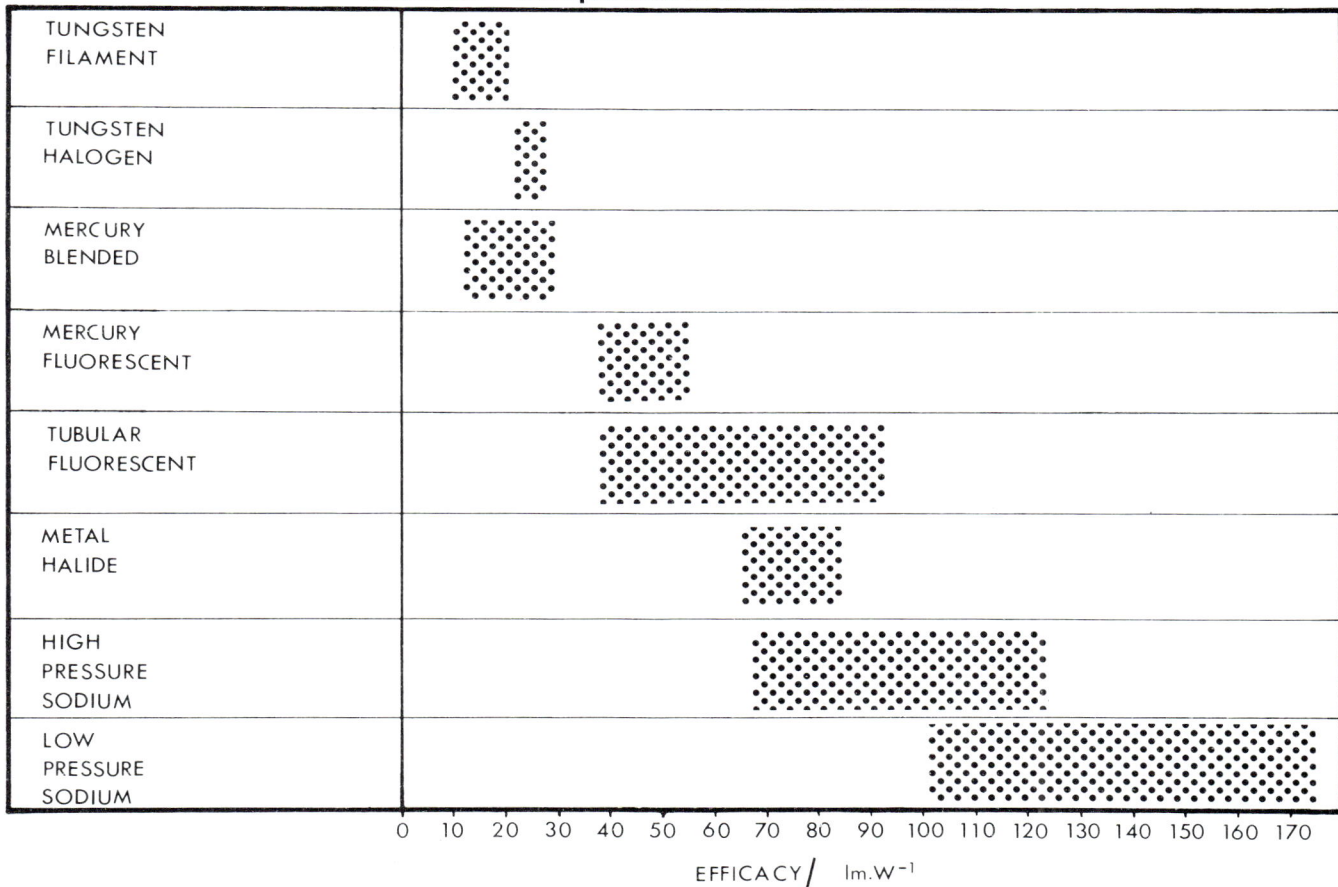

| | EFFICACY / lm.W⁻¹ |
|---|---|
| TUNGSTEN FILAMENT | ~10–20 |
| TUNGSTEN HALOGEN | ~20–25 |
| MERCURY BLENDED | ~10–25 |
| MERCURY FLUORESCENT | ~35–55 |
| TUBULAR FLUORESCENT | ~45–85 |
| METAL HALIDE | ~60–80 |
| HIGH PRESSURE SODIUM | ~65–120 |
| LOW PRESSURE SODIUM | ~100–175 |

EFFICACY / lm.W⁻¹

# 3.3 Luminaires

Luminaires can take many different forms, but all have to provide support, protection and electrical connection to the lamp. In addition luminaires have to be safe during installation and operation and be able to withstand the surrounding ambient conditions. The standard which covers most luminaires in the U.K. is BS 4533: Luminaires. It is suitable for use with luminaires containing tungsten filament, tubular fluorescent and other discharge lamps running on supply voltages not exceeding 1 kilovolt. It covers the electrical, mechanical and thermal aspects of safety. Luminaires should comply with BS 4533.

In BS 4533, luminaires are classified according to the type of protection against electric shock that they have, the degree of protection against ingress of dust or moisture they have, and according to the material of the supporting surface for which the luminaire is designed.

Table 3.4 lists the luminaire classes according to the type of protection against electric shock. Class 0 luminaires are not permitted in the U.K. by reason of the Electrical Equipment (Safety) Regulations and the Electricity (Factories Act) Special Regulations 1908 and 1944.

The degree of protection the luminaire provides against the ingress of dust and moisture is classified according to the Ingress Protection (IP) System. This system describes a luminaire by a two digit number, for example, IP 54. The first digit classifies the degree of protection a luminaire provides against the ingress of solid foreign bodies from fingers and tools, to fine dust. The second digit classifes the degree of protection a luminaire provides against the ingress of moisture. Table 3.5 lists the classes of these

**Table 3.4   Classification of luminaires according to the type of protection provided against electric shock (from BS 4533)**

| Class | Type of Protection | Symbol used to mark luminaires |
|---|---|---|
| 0* | A luminaire in which protection against electric shock relies upon basic insulation; this implies that there are no means for the connection of accessible conductive parts, if any, to the protective conductor in the fixed wiring of the installation, reliance in the event of a failure of the basic insulation being placed on the environment. | No symbol |
| I | A luminaire in which protection against electric shock does not rely on basic insulation only, but which includes an additional safety precaution in such a way that means are provided for the connection of accessible conductive parts to the protective (earthing) conductor in the fixed wiring of the installation in such a way that the accessible conductive parts cannot become live in the event of a failure of the basic insulation. | No symbol |
| II | A luminaire in which protection against electric shock does not rely on basic insulation only, but in which additional safety precautions such as double insulation or reinforced insulation are provided, there being no provision for protective earthing or reliance upon installation conditions. | ▱ |
| III | A luminaire in which protection against electric shock relies upon supply at safety extra low voltage (SELV) or in which voltages higher than SELV are not generated. The SELV is defined as a voltage which does not exceed 50 volts a.c., r.m.s. between conductors or between any conductor and earth in a circuit which is isolated from the supply mains by means such as a safety isolating transformer or converter with separate windings. | ◇III◇ |

*Class 0 luminaires are not permitted in the U.K.

**Table 3.5** **The degrees of protection against the ingress of solid bodies (first characteristic numeral) and moisture (second characteristic numeral) in the Ingress Protection (IP) System of luminaire classification**

| First character-istic numeral | Degree of protection | |
| :---: | :---: | :--- |
| | *Short description* | *Brief details of objects which will be 'excluded' from the luminaire* |
| 0 | Non-protected | No special protection |
| 1 | Protected against solid objects greater than 50 mm | A large surface of the body, such as a hand (but no protection against deliberate access). Solid objects exceeding 50 mm in diameter |
| 2 | Protected against solid objects greater than 12 mm | Fingers or similar objects not exceeding 80 mm in length. Solid objects exceeding 12mm in diameter |
| 3 | Protected against solid objects greater than 2.5 mm | Tools, wires, etc., of diameter or thickness greater than 2.5 mm. Solid objects exceeding 2.5 mm in diameter |
| 4 | Protected against solid objects greater than 1.0 mm | Wires or strips of thickness greater than 1.0 mm. Solid objects exceeding 1.0 mm in diameter |
| 5 | Dust-protected | Ingress of dust is not totally prevented but dust does not enter in sufficient quantity to interfere with satisfactory operation of the equipment |
| 6 | Dust-tight | No ingress of dust |

| Second character-istic numeral | Degree of protection | |
| :---: | :---: | :--- |
| | *Short description* | *Details of the type of protection provided by the luminaire* |
| 0 | Non-protected | No special protection |
| 1 | Protected against dripping water | Dripping water (vertically falling drops) shall have no harmful effect |
| 2 | Protected against dripping water when tilted up to 15° | Vertically dripping water shall have no harmful effect when the luminaire is tilted at any angle up to 15° from its normal position |
| 3 | Protected against spraying water | Water falling as a spray at an angle up to 60° from the vertical shall have no harmful effect |
| 4 | Protected against splashing water | Water splashed against the enclosure from any direction shall have no harmful effect |
| 5 | Protected against water jets | Water projected by a nozzle against the enclosure from any direction shall have no harmful effect |
| 6 | Protected against heavy seas | Water from heavy seas or water projected in powerful jets shall not enter the luminaire in harmful quantities |
| 7 | Protected against the effects of immersion | Ingress of water in a harmful quantity shall not be possible when the luminaire is immersed in water under defined conditions of pressure and time |
| 8 | Protected against submersion | The equipment is suitable for continuous submersion in water under conditions which shall be specified by the manufacturer. |

two digits. Table 3.6 lists the IP numbers which correspond to some commonly used descriptions of luminaire types and the symbols which may be used to mark the luminaires in addition to the IP number. Sometimes a third digit is used this refers to a French standard UTE C20 0 10 for impact testing.

Table 3.7 lists the classification of luminaires according to the material of the supporting surface for which the luminaire is designed.

Table 3.8 lists the information which should be distinctly and durably marked on a luminaire.

BS 4533 applies to most luminaires intended for use in neutral or hostile environments, (including luminaires with Type of Protection 'N' (Non-sparking)). It does not apply to many of the luminaires intended for use in hazardous environments, i.e. environments in which there is a risk of fire or explosion. For such applications there are different requirements so different standards and certification procedures apply. Detailed guidance on this topic can be found in the CIBS Application Guide: Lighting for hostile and hazardous environments.

**Table 3.6  Ingress Protection (IP) numbers corresponding to some commonly used descriptions of luminaire types and the symbols which may be used to mark a luminaire in addition to the IP number**

| Commonly used description of luminaire type | IP Number* | Symbol which may be used in addition to the IP classification number | |
|---|---|---|---|
| Ordinary | IP20** | no symbol | |
| Drip-proof | IPX1 | ◦ | (one drop) |
| Rain-proof | IPX3 | ▣ | (one drop in square) |
| Splash-proof | IPX4 | △ | (one drop in triangle) |
| Jet-proof | IPX5 | △△ | (two triangles with one drop in each) |
| Watertight (immersible) | IPX7 | ◦◦ | (two drops) |
| Pressure-watertight (submersible) | IPX8 | ◦◦ ₘ | (two drops followed by an indication of the maximum depth of submersion in metres) |
| Proof against 1 mm diameter probe | IP4X | no symbol | |
| Dust-proof | IP5X | ◈ | (a mesh without frame) |
| Dust-tight | IP6X | ◈ | (a mesh with frame) |

*Where X is used in an IP number in this Code, it indicates a missing characteristic numeral. However, on any luminaire, both appropriate characteristic numerals should be marked.

**Marking of IP 20 on ordinary luminaires is not required. In this context an ordinary luminaire is one without special protection against dirt or moisture.

**Table 3.7  Classification of luminaires according to the material of the supporting surface for which the luminaire is designed (from BS 4533)**

| Description of class | Symbol used to mark luminaires |
|---|---|
| Luminaires suitable for direct mounting only on non-combustible surfaces | No symbol – but a warning notice is required |
| Luminaires without built-in ballast or transformers, suitable for direct mounting on normally flammable surfaces | No symbol |
| Luminaires with built-in ballast or transformer suitable for direct mounting on normally flammable surfaces | ▽F |

**Table 3.8    Information which should be marked on luminaires (from BS 4533)**

1.  Mark of origin.

2.  Rated voltage(s) in volts. (Luminaires for tungsten filament lamps are only marked if the rated voltage is different from 250 volts).

3.  Rated maximum ambient temperature if other than 25°C ($t_a$ . . . °C).

4.  Symbol of class II or class III luminaire, where applicable (see table 3.4).

5.  Ingress Protection (IP) number, where applicable (see tables 3.5 and 3.6).

6.  Maker's model number or type reference.

7.  Rated wattage of the lamp(s) in watts. (Where the lamp wattage alone is insufficient, the number of lamps and the type shall also be given. Luminaires for tungsten filament lamps should be marked with the maximum rated wattage and number of lamps.)

8.  Symbol for luminaires with built-in ballast or transformers suitable for direct mounting on normally flammable surfaces, if applicable (see table 3.7).

9.  Information concerning special lamps, if applicable.

10. Symbol for luminaires using lamps of similar shape to 'cool beam' lamps where the use of a 'cool beam' lamp might impair safety, if applicable (   ).

11. Terminations to be clearly marked to identify which termination should be connected to the live side of the supply, where necessary for safety or to ensure satisfactory operation. Earthing terminators should be clearly indicated (   ).

12. Symbols for the minimum distance from lighted objects, for spotlights and the like, where applicable (   ).

In addition to the above markings, all details which are necessary to ensure proper installation, use and maintenance should be given either on the luminaire or on a built-in ballast or in the manufacturer's instructions provided with the luminaire.

### 3.3.2 Luminaire characteristics

Although meeting the requirements of BS 4533 is a common factor in luminaire design, it does little to limit the diversity of luminaires that are available. Luminaires vary in their construction, mounting position, distribution of light, the efficiency with which they provide light on the working plane, the extent to which they are likely to cause discomfort glare and the manner in which they light an interior. The most commonly occurring types of luminaire and their typical performance characteristics are given in table 3.9. The luminaire characteristics described are as follows:

*Mounting position.* Luminaires are usually either recessed into the ceiling (recessed), fixed on the ceiling (surface mounted), or suspended from the ceiling (pendant mounted). Some luminaires can be mounted in all three positions but most are only suitable for surface and/or pendant mounting. The mounting position is indicated by the letters R (recessed), S (surface), or P (pendant) as appropriate in column 2 of table 3.9.

*Polar curve shape.* The polar curve shape is a schematic illustration of the luminous intensity distribution of the luminaire. The luminous intensity distribution of the luminaire characterises the way in which the luminaire controls the light from the lamp. For linear luminaires, two curves, representing axial (A) and transverse (T) luminous intensity distributions are shown. For symmetrical luminaires a curve representing the average luminous intensity distribution is given. The polar curve shape for each luminaire type is given in column 3 of table 3.9.

*Nominal spacing/mounting height ratio (SHR NOM).* When designing a uniform lighting installation using the lumen method (see Part 4: Design) conformance with the uniformity criterion is ensured by limiting the spacing between the luminaire centres. The maximum spacing that is allowable is determined by the luminous intensity distribution of the luminaire and its mounting height. Therefore the spacing allowable for each luminaire type is expressed as a ratio of the spacing to the mounting height. Typical nominal values of spacing/mounting height ratio are given for each luminaire type in column 4 of table 3.9. (For a more detailed consideration of spacing/ mounting height ratio, see section 4.5.3.5.)

*Utilisation factor.* The efficiency of a lamp/luminaire combination when used to provide uniform lighting in a particular room can be expressed by the utilisation factor. The utilisation factor is the luminous flux which reaches the working plane expressed as a ratio of the luminous flux emitted by the lamp. For each luminaire type, the utilisation factor will vary with the efficiency, distribution and spacing of the luminaires as well as with the room proportions and the reflectance of the room surfaces. Thus for any single luminaire type, a range of utilisation factors will be obtained. An indication of the range of utilisation factors likely to occur in rooms of different room index, with low and high surface reflectances, is given for a regular array at the nominal spacing of a typical luminaire of each type, in the form shown below.

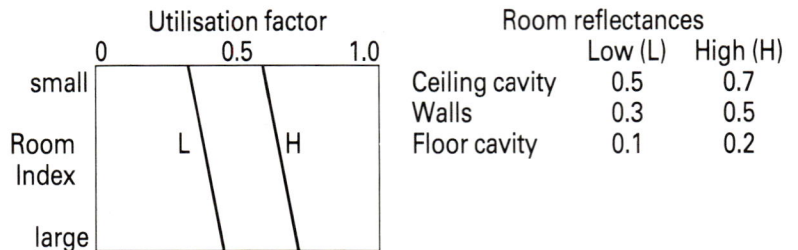

This information is given in column 5 of table 3.9.

*Glare Index.* The luminous intensity distribution of a luminaire and its projected area are important factors in determining the Glare Index of the installation. The Glare Index produced by a regular array of luminaires of any particular type will also vary with the room proportions, the room surface reflectances and the room and luminaire orientation relative to the line of sight. Column 6 of table 3.9 gives an indication of the range of Glare Indices likely to occur in square rooms of different areas with low and high surface reflectances, for a regular array of each luminaire type, in the form shown below. For linear luminaires solid lines indicate limits of endwise viewing, broken lines indicate limits for crosswise viewing.

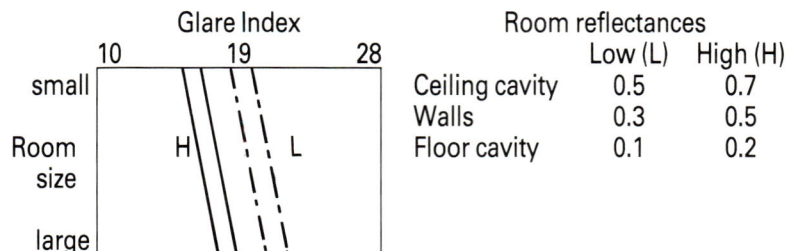

*Room surface brightness.* The luminous flux distribution of a luminaire determines to some extent the relative brightness of walls and ceilings. For a regular array of luminaires of each type, the pattern of brightness of walls and ceilings for rooms with average surface reflectances are indicated in column 7 of table 3.9. The key to the diagram is shown below.

*Luminaires*

| | | |
|---|---|---|
| Bright Ceiling | Bright Ceiling | Bright Ceiling |
| Dull Wall | Medium Wall | Bright Wall |
| Medium Ceiling | Medium Ceiling | Medium Ceiling |
| Dull Wall | Medium Wall | Bright Wall |
| Dull Ceiling | Dull Ceiling | Dull Ceiling |
| Dull Wall | Medium Wall | Bright Wall |

Increasing ceiling brightness ↑

Increasing wall brightness →

**3.3.3 Caution**

It must be emphasised that table 3.9 is intended only as an overview; it is not suitable for lighting design. However, it can guide the user as to the choice and type of luminaire needed for a particular set of lighting criteria. The information in table 3.9 is not a substitute for the manufacturer's data. There are many forms of luminaire other than those displayed and the same type of luminaire can show considerable variation in its characteristics. For accurate information, the manufacturers should always be consulted. Photometric data supplied by manufacturers should have been obtained from photometric measurements according to BS 5225 and calculated according to CIBS Technical Memorandum 5: The calculation and use of utilisation factors.

# Table 3.9    Luminaire Characteristic Charts

| Luminaire | Mounting | Polar Curve | SHR NOM | Utilisation Factor | Glare Index | Surface Brightness |
|---|---|---|---|---|---|---|
| Batten | SP | | 1.75 | | | |
| Batten | SP | | 1.75 | | | |
| Batten | SP | | 1.75 | | | |
| Slim opal diffuser | SP | | 1.75 | | | |
| Slim prismatic | SP | | 1.50 | | | |
| Medium single prismatic | SP | | 1.75 | | | |
| Wide twin prismatic | SP | | 1.75 | | | |

| Luminaire | Mounting | Polar Curve | SHR NOM | Utilisation Factor | Glare Index | Surface Brightness |
|---|---|---|---|---|---|---|
| Wide twin opal diffuser | SP | | 1.75 | | | |
| Slotted white trough reflector | SP | | 1.75 | | | |
| Slotted white trough reflector | SP | | 1.5 | | | |
| Elliptical aluminium reflector with transverse blades | RSP | | 1.0 | | | |
| Faceted aluminium batwing reflector with transverse blades | RSP | | 1.75 | | | |
| Metalized parabolic cell louvre | RSP | | 1.00 | | Below 10 | |
| Recessed modular with flat sheet prismatic lens | R | | 1.50 | | | |

**Table 3.9** Continued

**WARNING: NOT SUITABLE FOR LIGHTING DESIGN**

| Luminaire | Mounting | Polar Curve | SHR NOM | Utilisation Factor | Glare Index | Surface Brightness |
|---|---|---|---|---|---|---|
| Recessed modular with opal dished diffuser | R | | 1.50 | | | |
| Recessed modular with 45° cut off white cell louvre | R | | 1.25 | | | |
| Opal sides prismatic base | SP | | 1.50 | | | |
| Opal sides prismatic base | SP | | 1.50 | | | |
| Opal sides and base | SP | | 1.50 | | | |
| Opal sides and base | SP | | 1.50 | | | |
| Industrial clear patterned | SP | | 1.75 | | | |

**Table 3.9**   Continued

**WARNING: NOT SUITABLE FOR LIGHTING DESIGN**

| *Luminaire* | *Mount-ing* | *Polar Curve* | *SHR NOM* | *Utilisation Factor* | *Glare Index* | *Surface Brightness* |
|---|---|---|---|---|---|---|
| Industrial clear patterned | SP | | 1.50 | | | |
| High bay concentrating reflector | P | | 0.75 | | | |
| High bay medium spread reflector | P | | 1.25 | | | |
| High bay batwing reflector | P | | 1.75 | | | |
| High bay diffuser reflector | P | | 1.50 | | | |
| Reflector lamp with skirt reflector/shield | P | | 1.25 | | | |
| Box with batwing facetted reflector | RSP | | 1.75 | | | |

**Table 3.9** Continued

| Luminaire | Mounting | Polar Curve | SHR NOM | Utilisation Factor | Glare Index | Surface Brightness |
|---|---|---|---|---|---|---|
| Box with prismatic base | RSP | | 1.25 | | | |
| Box with opal base | RSP | | 1.50 | | | |
| Downlight, narrow beam with Par 38 lamp | RSP | | 0.25 | | Below 10 | |
| Downlight, narrow beam reflector lamp | RSP | | 0.50 | | Below 10 | |
| Downlight GLS lamp plus reflector | RSP | | 1.25 | | | |
| Opal bowl | S | | 1.50 | | | |
| Opal sphere | SP | | 1.75 | | | |

**Table 3.9**   Continued

**WARNING: NOT SUITABLE FOR LIGHTING DESIGN**

| Luminaire | Mount-ing | Polar Curve | SHR NOM | Utilisation Factor | Glare Index | Surface Brightness |
|-----------|-----------|-------------|---------|--------------------|-------------|--------------------|
| Opal cylinder | SP | | 1.75 | | | |
| Circular fluorescent, prismatic bowl | SP | | 1.25 | | | |
| Circular fluorescent, opal bowl | SP | | 1.50 | | | |

# 3.4 Control systems

### 3.4.1 The function of control systems

Control systems are an inherent part of any lighting installation. They can take many forms, varying from a simple wall switch to being a part of a sophisticated microprocessor-controlled, building management system. Whatever the method used, the aim of a control system is always to ensure that the lighting system is only operating when it is required, and that when it is, it is operating in the required state. The aim of most control systems is to vary the light output of the installation, either by switching or by dimming the lamps (see section 4.4.4.1).

### 3.4.2 Switching

In principle, all light sources can be switched but the light output that is immediately available on switch on and the interval necessary between switch off and switch on varies with lamp type (see table 3.2). Switching can be achieved by a number of different methods. The simplest is the manual switch. Remote switches which use an infrared transmitter and a receiver on the luminaire are also available. Both these forms of switching require human initiative. Alternative forms of switching operate without human intervention. Lamps can be switched by time switches or in response to the availability of daylight or the occupation of the interior. Photocells are used to sense the level of daylight available in an interior, whilst sensors of noise level, movement and reflected radiation have all been used to detect people's presence in an interior.

One particular aspect of switching which has limited its use in the past has been the difficulty of switching individual or small groups of luminaires without excessive investment in wiring. Recent developments in electronics have made it possible to send switching signals by low voltage wiring or by high frequency transmission pulses over the existing supply wiring. Further, logic circuitry, now exists which allows individual luminaires to respond in one of several different ways. Such systems provide great flexibility in the way the lighting installation can be used.

### 3.4.3 Dimming

Whenever the ability to steadily diminish the illuminance in a room is desirable, dimming is required. Tungsten filament lamps can be readily dimmed. Not all discharge lamps can be dimmed and those that can, such as tubular fluorescent lamps, need special control gear. Dimming lamps reduces the energy consumed by the lamp, but not necessarily in proportion to the light output, and usually changes its colour properties. Dimmers can be controlled manually or automatically in response to daylight availability. Many of the electronic developments mentioned in relation to switching can also be associated with dimming.

# 3.5 Maintenance of lighting equipment

### 3.5.1 Introduction

Maintenance of lighting systems keeps the performance of the system within the design limits, promotes safety, and, if considered at the design stage, can help to minimise the electrical load and capital costs. Maintenance includes replacement of failed or deteriorated lamps and control gear, and the cleaning of luminaires and room surfaces at suitable intervals.

Lighting systems need maintaining because without it they deteriorate. The light output from lamps decreases with time of operation until the lamp fails. Different lamp types deteriorate at different rates. Further, dirt deposition will occur on lamps, luminaires and room surfaces. Figure 3.1 shows the changes in the illuminance produced over time by an installation of tubular fluorescent lamps, with various combinations of lamp replacement and luminaire cleaning intervals. Different lamp types in different luminaires in different locations will produce a different pattern but the underlying processes are the same.

## 3.5.2 Lamp replacement

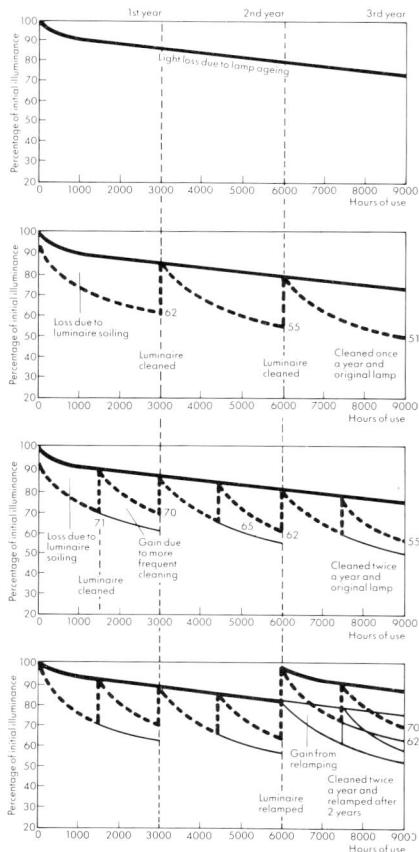

**Fig 3.1** Changes in illuminance with time for different maintenance schedules

There are two factors which need to be considered when determining the timing of lamp replacement, the change in light output and the probability of lamp failure. The relative weight given to these two factors depends on the lamp type. Tungsten filament, tungsten halogen and blended high pressure mercury tungsten discharge lamps, all of which have a filament, usually fail before the decline in light output becomes significant. Therefore the replacement time for these lamps is determined by the probability of lamp failure alone. All the other light sources conventionally used in interiors show a significant reduction in light output before a large proportion fail. Therefore, for these lamps, both the decline in light output and the probability of lamp failure are important in determining the lamp replacement time. Frequently it is desirable to replace such lamps even though they are still operating electrically, simply because the light output has fallen to an uneconomic level.

For the majority of installations the most sensible procedure is to replace all the lamps at planned intervals. This procedure, which is known as group replacement, has visual, electrical and financial advantages over the alternative of replacing individual lamps as they fail. Visually, group replacement ensures that the installation maintains a uniform appearance. Electrically, group replacement reduces the risk of damage to the control gear caused by lamps nearing the end of their electrical life. Financially, by arranging that the lamp replacement is associated with luminaire cleaning and doing it at a time when it will cause the minimum of disturbance to the activities in the interior, the cost of lamp replacement can be minimised. Group replacement is an appropriate procedure for routine maintenance. However, in any large installation, a few lamps can be expected to fail prematurely. These lamps should be replaced promptly on an individual basis.

No matter whether lamps are replaced individually or in a group a decision has to be made about the replacement light source. As light source development proceeds there is a temptation to replace one light source with another which is superficially similar but of higher luminous efficacy. If this course of action is attempted great care should be taken to establish that the replacement light source and the existing control gear are compatible; physically and electrically. Before replacing any discharge light source with another of a different type, or the same type but from a different manufacturer, advice on compatibility should be sought from the manufacturers.

The timing and nature of lamp replacement is usually a matter of economic and managerial judgement and may well be determined by factors other than those directly related to the lighting. The proposed lamp replacement procedure should be considered during the design of the installation.

## 3.5.3 Luminaire cleaning interval

The rate at which dirt is deposited on and in a luminaire depends on the amount and composition of the dirt in the atmosphere, and on the type of luminaire. Over the same period and in the same location dustproof (IP 5X) and dust tight (IP 6X) luminaires and open reflectors with slots in the top will collect less dirt than louvred luminaires with closed tops, or luminaires with unsealed diffusers. This last type frequently act as dust traps.

For particularly dirty atmospheres or where access is difficult it is usually best to have either dustproof (IP 5X) or dust tight (IP 6X) luminaires, ventilated luminaires which are designed to use air currents to keep them clean, or lamps with internal reflectors. If cleaning is to be done in situ by water jets then luminaires suitably protected from moisture penetration must be used (IP X5). It is important to note that even the most protected luminaires, for example, dust tight (IP 6X) luminaires, will collect dirt on their external surfaces. Therefore even these luminaires will need cleaning regularly.

The appropriate cleaning interval for luminaires and the lamps they contain is again an economic and managerial question. The factors that need to be considered are the cost and convenience of cleaning at a particular time and the prevailing efficiency of the installation. As a general guide, luminaires should be cleaned at least once a year but for some locations this will not be sufficient. Appendix 7 gives some representative luminaire light output depreciation curves for different luminaire types and atmospheric conditions. From such curves it is possible to establish a suitable cleaning interval. It is usually advantageous to co-ordinate luminaire cleaning with lamp replacement if the latter is required.

### 3.5.4 Room surface cleaning interval

The Factories Act 1961 stipulates that room surfaces in factories should be cleaned and redecorated regularly, but regular cleaning is important in all buildings if a dirty appearance is to be avoided. Regular cleaning is particularly important where light reflected from the room surfaces makes an important contribution to the lighting of the interior, e.g. where daylight from the side windows is used or where an indirect lighting installation is present.

### 3.5.5 Design aspects

The maintenance procedures for a lighting installation should be considered at the design stage. Three aspects are particularly important. The first is the light loss factor to be used in the calculation of the number of lamps and luminaires needed to provide the required conditions (see appendix 7). The closer the light loss factor is to unity, the smaller the number of lamps and luminaires which will be needed to provide the desired conditions. However, it must be emphasised that a light loss factor close to unity implies a commitment to regular and frequent maintenance. Unless this commitment is fulfilled the installation will not meet the design conditions.

The second aspect is the practical one of access and handling. Good maintenance will only occur if access to the lighting installation is safe and easy, and the lighting equipment is simple to handle.

The third aspect is equipment selection. The dirtier the atmosphere where the installation is to operate, the more important it is to select equipment which is resistant to dirt deposition.

### 3.5.6 Practical aspects

A wide range of different materials are used in luminaires. Table 3.10 summarises the most suitable cleaning methods for use with these different materials.

Different lighting installations call for different levels of skill from the people doing the maintenance. For completely conventional installations only the most basic knowledge is necessary but where luminaires with special properties, e.g. dustproof (IP 5X), jetproof (IP X5), are used, considerable knowledge and care is required from the operator doing the cleaning and reassembly. Similarly, where complex control systems form a part of the installation, the maintenance operator will need to understand the operation of the system and the consequences of any changes made. For all maintenance work the operators should be aware of the basic electrical and mechanical safety aspects of the work.

**Table 3.10   Suitable cleaning methods for lighting maintenance**

| Materials | | Cleaning method |
|---|---|---|
| Aluminium reflectors | (a) | Clean surfaces with a damp cloth or sponge. A detergent may be used but this should be rinsed away and the surface dried/polished with a soft cloth. |
| | (b) | Where there is obvious staining or contamination a metal polish or paste such as 'Brasso' may be used followed by a cleaning operation as in (a). |
| | (c) | Badly tarnished or corroded surfaces cannot be restored. Reflectivity may be improved by removing the tarnished oxide with chromic acid but corrosion and abrasion resistance will be affected. It is better at this stage to replace the reflectors. |
| Plastics (including G.R.P.) | (d) | Clean as (a) above but some dust accumulation can be expected as a result of static charge. It is better to use an antistatic cleaning fluid in place of water and the detergent but these should not be used for components under continuous stress, e.g. under the tension of springs. Organic solvents should also be avoided. |
| | (e) | Aged or yellowed surfaces can be improved by using an abrasive (on the yellowed surface) but again stressed components should not be treated in this way. Restore gloss as in (b). |
| Paints | (f) | Clean as in (a). Aged or yellowed surfaces must not be abraded to restore colour. |
| Glass and Vitreous Enamel | (g) | Clean as in (a) or by utilising commercially available cleaning/polishing products. Oily substances such as paraffin will give 'rainbow patterns' and should be avoided. |
| Galvanised surfaces | (h) | Clean as in (a) but components must be dried. 'White corrosion films' may be removed by *light* abrasion with fine (600 grit) emery paper or other abrasive compounds but rusting will then occur at an earlier stage. |

## 4.1 Introduction

Lighting design is a complex process and no hard and fast rules can be devised which will suit all design problems or every designer. Nevertheless, the following design approach represents reasonable practice and will give guidance to less experienced designers.

A flowchart of the overall process is given in figure 4.1, and each stage is detailed in the following sections.

## 4.2 Objectives

The first stage in planning any lighting installation is to establish the lighting design objectives. Care and time expended on this is well invested, because the objectives guide the decisions in all the other stages of the design process. Establishing the objectives for a design is not the same as reading Part 2 of this Code and deciding upon a suitable illuminance. Rather, it is a matter of deciding what the lighting is for. The lighting objectives can be considered in three parts.

**4.2.1 Safety**

The lighting must be safe in itself and must allow the occupants to work and move about safely. These are not only primary objectives but also statutory obligations. It is, therefore, necessary to identify any hazards present and to consider the most appropriate form of emergency lighting (see section 4.5.6).

**4.2.2 Visual tasks**

The type of work which takes place in the interior will define the nature and variety of the visual tasks. An analysis of the visual tasks (there is rarely just one) in terms of size, contrast, duration, need for colour discrimination and so on, is essential to establish the quantity and quality of the lighting required to achieve satisfactory visual conditions.

The designer should be wary of general descriptions. A general office, for example, can, at one extreme, have occupants whose job it is to answer the telephone and, as a result, the visual tasks may be quite simple; whilst, at the other extreme the occupants may have to transcribe text, handwritten in pencil, onto terminals equipped with VDUs, and this presents a complex set of visual tasks. In addition to establishing the nature of the tasks done in an interior, it is also necessary to identify the positions where the tasks occur and the planes on which the tasks lie. This information is essential if lighting matched to the tasks is to be provided.

**4.2.3 Appearance and character**

The lighting of a space will affect its character, and the character of objects within it. It is, therefore, necessary to establish what mood or atmosphere is to be created. This is not a luxury to be reserved only for prestige offices, places of entertainment, and the like, but should be considered in all designs, even for those areas where it will be given less importance than other factors.

**4.2.4 Priorities and constraints**

When establishing the objectives, it is important to differentiate between those which are essential and those which are desirable. It is also important at this stage to establish both the design objectives and the design constraints.

Often the most obvious and the most important constraint is financial. Obviously everyone wants to spend the minimum possible, but different owners will spend differing amounts, for otherwise similar areas, according to their own valuation of the final result. For this reason, it is important to establish the financial constraints. These will temper and modify the importance of the various design objectives, but should be opposed if they suppress any of the essential objectives.

Ideally, it should be possible to consider both capital expenditure and running costs to achieve the most economical scheme. In some circumstances this does not happen, and only the capital cost is considered, because a second system of budget control applies to the running cost. Although this is a highly unsatisfactory approach and should be resisted, it may be forced upon the designer. More usually it is possible to relate capital and running costs to establish the lowest overall investment. The normal method of doing this is to calculate fixed and variable costs, allowing for depreciation, interest and inflation, over a fixed term.

There are many other constraints which may affect the design objectives, such as energy consumption, environmental considerations (which may limit the range of acceptable luminaires), physical problems of access, and so on. These constraints must be recognised at the objectives stage of the design.

**Figure 4.1   Lighting design flowchart**

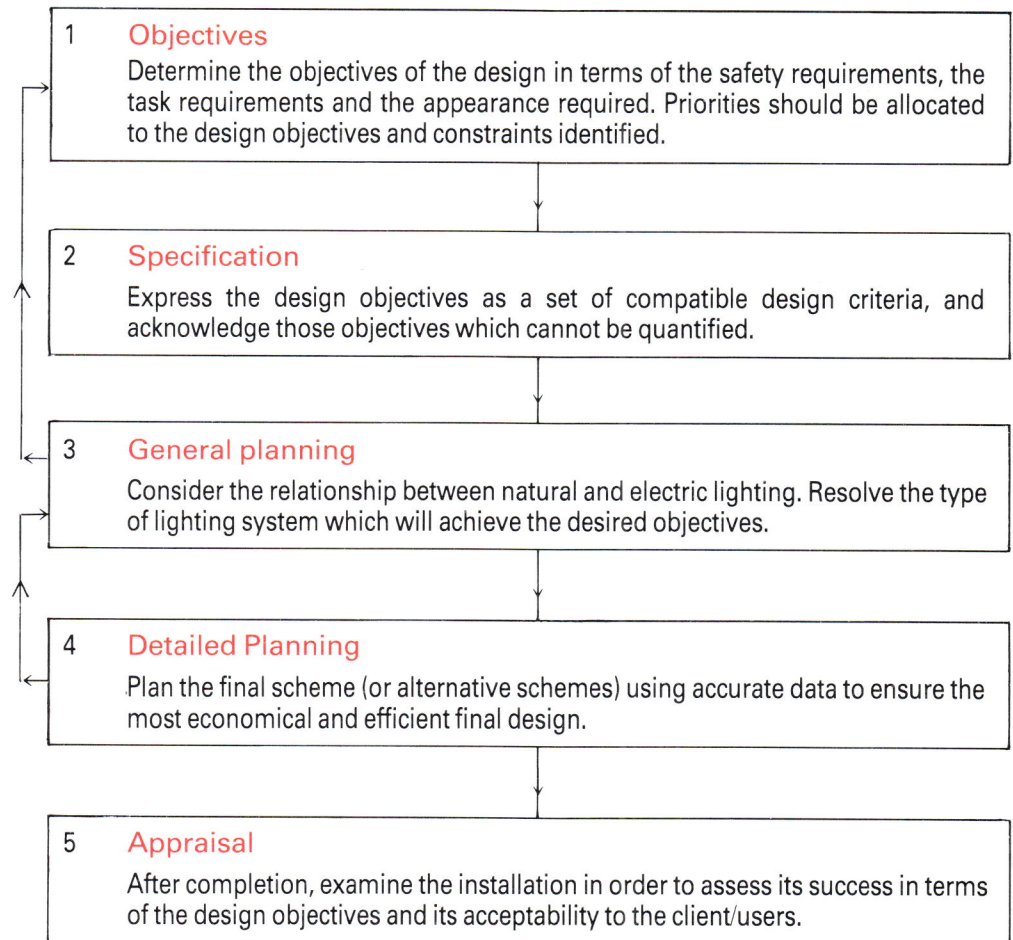

| 1 | **Objectives** |
|---|---|
| | Determine the objectives of the design in terms of the safety requirements, the task requirements and the appearance required. Priorities should be allocated to the design objectives and constraints identified. |

| 2 | **Specification** |
|---|---|
| | Express the design objectives as a set of compatible design criteria, and acknowledge those objectives which cannot be quantified. |

| 3 | **General planning** |
|---|---|
| | Consider the relationship between natural and electric lighting. Resolve the type of lighting system which will achieve the desired objectives. |

| 4 | **Detailed Planning** |
|---|---|
| | Plan the final scheme (or alternative schemes) using accurate data to ensure the most economical and efficient final design. |

| 5 | **Appraisal** |
|---|---|
| | After completion, examine the installation in order to assess its success in terms of the design objectives and its acceptability to the client/users. |

# 4.3  Specification

Once the lighting objectives have been defined, they must be expressed in a suitable form. Not all design objectives can be expressed as measurable quantities. For example, the need to make an environment appear 'prestigious', 'efficient' or 'vibrant' cannot be quantified. Furthermore, although many objectives can be expressed in physical terms, suitable design techniques may not exist or may be too cumbersome. For example, obstruction losses and contrast rendering factors (see appendix 6), are two quantities that are difficult to calculate and predict accurately. This does not mean that the objectives represented by these terms should be ignored, but that experience and judgement may have to replace calculation.

Lighting designers have a responsibility to ensure that lighting is not liable to cause injury to the health of occupants (appendix 8). Bad lighting can contribute towards accidents or result in inadequate working conditions. The designer must always take due note of statutory instruments that affect lighting conditions, such as the Health and Safety at Work etc. Act 1974, Factories Act 1961, and so on. Most of these demand that lighting shall be both sufficient and suitable. Sufficiency is normally taken to be related to the quantity of illumination (illuminance) on the tasks and for safe movement, whilst suitability covers discomfort and disability glare, spectral distributions, veiling reflections, shadows, and so on.

Legislation is concerned with what is essential. This is less onerous than the recommendations in this Interior Lighting Code, which are concerned with good practice.

A full specification can be established by reference to other parts of this Code and by taking the design objectives into account. The factors which can be specified numerically are: *Financial Budget* (see section 4.5.1.1), *Energy Budget* (see sections 2.5 and 4.5.1.3), *Design Service Illuminance* (see sections 2.3, 2.4.2 and 1.3), *Uniformity* (see sections 2.4.3 and 1.3.4), *Modelling, vector/scalar ratio* (see sections 2.4.4 and 1.4), *Reflectances and Colours of Surfaces* (see sections 2.4.5, 1.5 and Appendix 3), *Illuminance Ratios* (see section 2.4.3), *Light Source Colour* (see sections 2.4.6 and 1.6.1), *Colour Rendering Requirement* (see sections 2.4.6, 1.6.2 and Appendix 4), *Glare* (see sections 2.3, 2.4.7, 1.7 and Appendix 5), *Contrast Rendering* (see section 1.8 and Appendix 6), *Run-up/Restrike Time* (see section 3.2.2).

Care must be taken to ensure that those objectives which cannot be quantified are not overlooked.

# 4.4  General Planning

When the design specification has been established the purpose of the remaining stages of design is to translate these physical requirements into the best possible solutions, with the intention of meeting the original objectives. The designer should never lose sight of the fact that the aim is to meet the original objectives, and that the specification is only a stepping stone in this process, and not an end in itself. Indeed, if it proves difficult to plan an installation which meets the design specification it may be necessary to reassess the original objectives. There are no hard and fast rules about how to plan a lighting installation, and experience and judgement will usually dominate the planning process. Nevertheless, the planning stages can be divided into *general planning* and *detailed planning* (see section 4.5).

At the general planning stage, the designer aims to establish whether the original objectives are viable, and resolve what type of design can be employed to satisfy these objectives. The first stage in the general planning of a lighting installation is to consider the interior to be lit, its proportions, its contents, and most importantly the daylight available.

**4.4.1  Daylight**

In the past, the chief aim of most windows has been to provide light. Today, other environmental factors may demand equal consideration – solar gain in summer, fabric heat loss in winter, natural ventilation, the entry of noise and dirt from outside, the view in and out, the composition of the architectural facade. Most of these aspects of window design will be considered in a forthcoming CIBS publication.

This Code seeks to offer guidance to a lighting engineer confronted by an existing window or rooflight. The questions which need to be considered in these circumstances are, how should the electric lighting relate to the natural lighting, and does the window or rooflight need additional visual or thermal protection.

### 4.4.1.1 The relation between natural and electric lighting

Electric lighting is usually planned as if daylight did not exist. However, natural lighting may well suggest the form and especially the control system of the electric lighting. For this reason, every lighting designer needs some knowledge of daylight prediction.

### 4.4.1.2 Average daylight factors

In temperate climates, the extent to which daylight is available at a position in an interior is conventionally expressed as a 'daylight factor'. This is the illuminance at a point on a plane in an interior due to light received directly or indirectly from a sky of known or assumed luminance distribution expressed as a percentage of the illuminance on a horizontal plane due to an unobstructed hemisphere of the same sky. The sky usually assumed is the overcast sky specified by the CIE (see appendix 9). The average daylight factor on a horizontal reference plane in an empty interior is given approximately by the following expression:

$$\text{average daylight factor} = \frac{T\,W\,\theta}{A(1-R^2)} \text{ per cent}$$

where $T$ = transmittance of glazing material, expressed as a decimal
$W$ = nett area of glazing (m$^2$)
$\theta$ = angle (degrees) in vertical plane, subtended by sky visible from the centre of a window or a rooflight – (Fig 4.2)
$A$ = total area of indoor surfaces: ceiling + floor + walls, including windows or rooflights (m$^2$)
$R$ = area weighted average reflectance of all indoor surfaces, including windows or rooflights

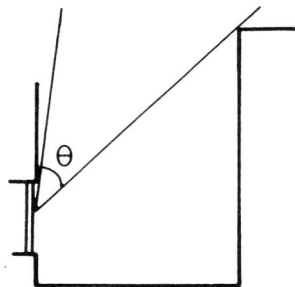

**Fig 4.2 (a)** θ is the angle subtended in a vertical plane normal to the window, by sky visible from the centre of the window.

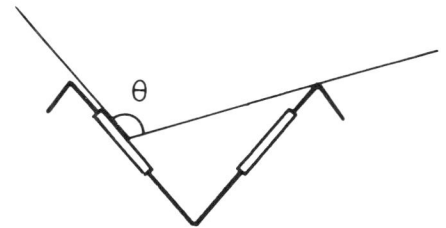

**Fig 4.2 (b)** θ is the angle subtended in a vertical plane normal to the rooflight, by sky visible from the centre of a rooflight in a shed roof

The average daylight factor is a suitable criterion for shallow side-lit interiors and interiors lit by rooflights. The average daylight factor is not an appropriate criterion for the deep side-lit interiors discussed in section 4.4.1.3, the daylight factor distribution being clearly uneven. The equation for average daylight factor does not apply to such an interior. Subject to these reservations, it appears that when the average daylight factor is 5 per cent or more, an interior will generally look well day-lit. When the average daylight factor is less than 2 per cent, the interior will not be perceived as well day-lit and electric lighting may be in constant use throughout the day.

When the average daylight factor exceeds 5 per cent in a building which is used mainly during the day, electricity consumption for lighting should be too small to justify elaborate control systems on economic grounds, provided that switches are sensibly located. Lamps, which will be used mainly at night, should be of the 'warm or intermediate' CCT class (Table 1.1).

When the average daylight factor is between 2 per cent and 5 per cent, the electric lighting should be planned to take full advantage of available daylight. Localised or local lighting may be particularly advantageous, using daylight to provide the general surround lighting. Lamps with a correlated colour temperature in the upper half of the intermediate CCT class blend well with daylight.

When the average daylight factor is below 2 per cent, supplementary electric lighting will be needed almost permanently.

Average daylight factor will often given the designer sufficient information on which to base decisions on the relationship between natural and electric lighting. However, where more detailed information on daylight factors is necessary, the methods for point-by-point daylight factor prediction described in Appendix 9 can be used.

## 4.4.1.3. Daylight in deep side-lit interiors

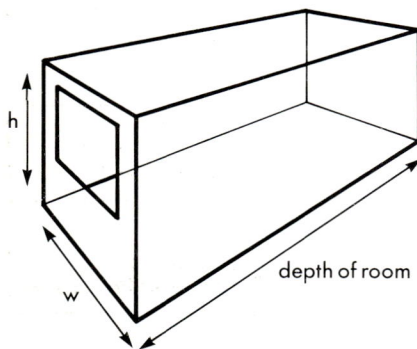

Fig 4.3 Limiting room proportions

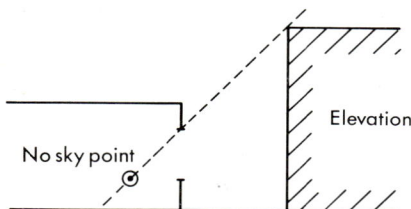

Fig 4.4 The no-sky line, which is the locus of points beyond which it is impossible for an occupant to see any part of the sky

Some rooms appear to be fairly evenly daylit. These include many roof-lit or clerestory-lit interiors and many shallow side-lit rooms.

In a deep side-lit room, however, the lighting in the depth of the interior may look very dull when compared with the lighting just inside the window. This is likely to occur when the depth of the room, from window to back wall, is greater than the limiting depth calculated from the expression:

$$D = \frac{2\,w\,h}{(h+w)(1-R_B)}$$

where  $D$  = limiting depth
$w$  = width of room measured from side to side, parallel to window, (Fig 4.3)
$h$  = height of window head above floor (Fig 4.3)
$R_B$  = area-weighted average reflectance of surfaces in the half of the room remote from the window

If the room is lit from two opposite sides, the limiting depth, calculated as above, may be doubled.

Further, a side-lit room having smaller depth than the limiting value will still look unevenly lit if part of the working surface lies behind the no-sky line (Fig 4.4).

This has repercussions for electric lighting. If the no-sky line impinges seriously on the working plane or if the room depth exceeds the limiting value defined above, the electric lighting is likely to be in continuous use whenever the room is occupied, whatever the shape, size or position of the window.

## 4.4.1.4 Protection from glare and solar gain

Windows and rooflights can become sources of glare because they allow a direct view of the sun or a bright sky. In practice the strongest complaints are associated with direct sunlight. Therefore, the best strategy for reducing discomfort from windows and rooflights is to concentrate on controlling solar penetration; this will reduce sky glare too.

The appropriate form of solar protection will depend on whether the discomfort is mainly thermal or visual; if both, some combination of screening measures may be needed. Protection may be fixed or movable. Although fixed baffles can simplify maintenance and ensure a tidier facade, movable protection is generally preferable in the United Kingdom as it causes less obstruction to daylight and view on dull days. The flow charts in figures 4.5a and 4.5b offer a systematic procedure for selecting solar protection for windows and rooflights.

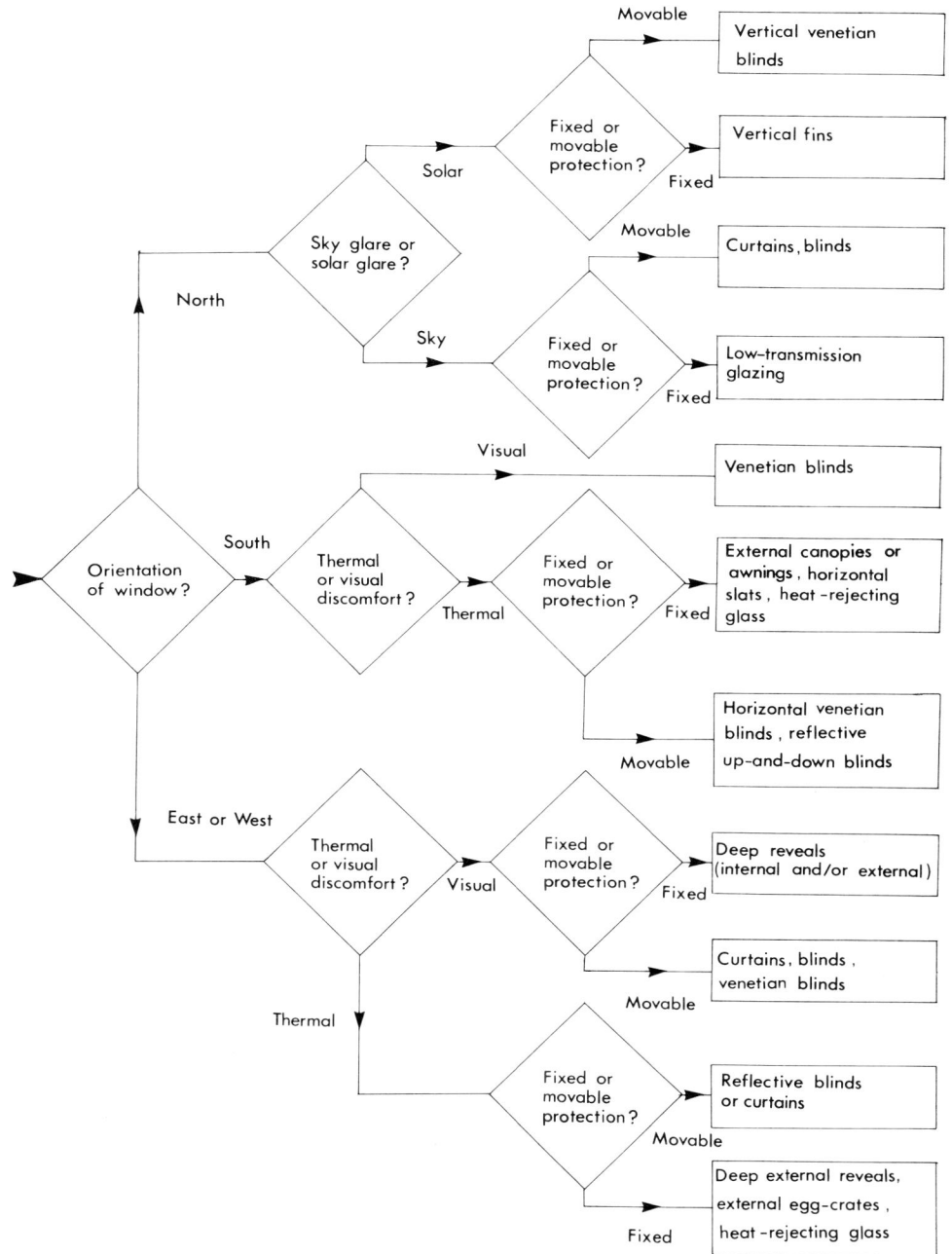

**Fig 4.5 (a)** A flowchart for selecting solar protection for windows

### 4.4.2 Choice of electric lighting systems
#### 4.4.2.1 General lighting

Lighting systems which provide an approximately uniform illuminance over the whole working plane are called general lighting systems (see Fig 4.6a). The luminaires are normally arranged in a regular layout. The appearance of the installation is usually tidy but may be rather bland. General lighting is simple to plan using the lumen method (see section 4.5.3.6) and requires no co-ordination with task locations. The greatest advantage of such systems is that they permit complete flexibility of task location.

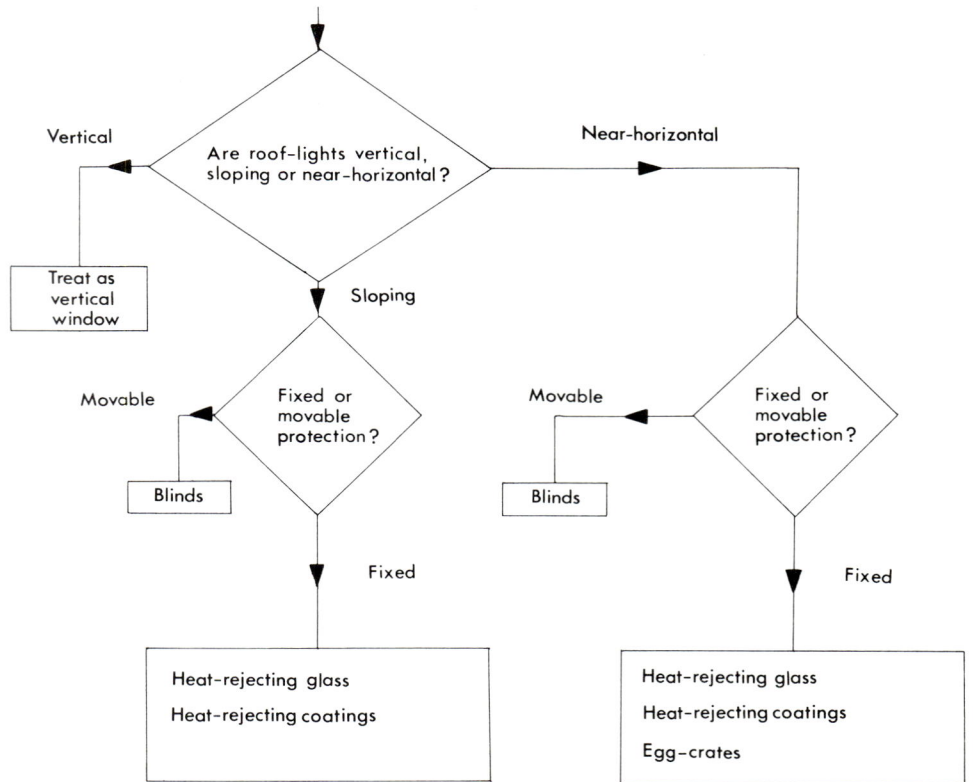

Fig 4.5 (b)  A flowchart for selecting window protection for rooflights

Fig 4.6 (a)  A general lighting system employs a regular array of luminaires to provide a uniform illuminance across the working plane

Fig 4.6 (b)  A localised lighting system uses luminaires located adjacent to the work stations to provide the required task illuminance. The necessary ambient illuminance in the surrounding areas is provided by additional luminaires as required

Fig 4.6 (c)  A local lighting system employs luminaires, located at the workstation, to provide the necessary task illuminance. A general lighting scheme is used to provide the ambient illuminance for the main area

The major disadvantage of general lighting systems is that energy may be wasted illuminating the whole area to the level needed for the most critical tasks. Energy could be saved by providing the necessary illuminance over only the task areas and using a lower ambient level for circulation and other non-critical tasks.

## 4.4.2.2 Localised lighting

Localised lighting systems (see Fig 4.6b) employ an arrangement of luminaires designed to provide the required service illuminance on work areas together with a lower illuminance for the other areas. The illuminance on the other areas should not be less than one-third of the illuminance on the work areas.

Considerable care must be taken to co-ordinate the lighting layout to task positions and orientation. The system can be inflexible and correct information is essential at the design stage. Changes in the work layout can seriously impair a localised system, although uplighters and other easily relocatable or switchable systems can overcome these problems.

Localised systems normally consume less energy than general lighting systems unless a high proportion of the area is occupied by work stations. This should be confirmed by specific calculations. Localised systems may require more maintenance than general lighting systems.

## 4.4.2.3 Local lighting

Local lighting provides illumination only over the small area occupied by the task and its immediate surroundings (Fig 4.6c). A general lighting system must be installed to provide sufficient ambient illumination for circulation and non-critical tasks. This is then supplemented by the local lighting system to achieve the necessary design service illuminance on tasks. The general surround illuminance should not be less than one-third of the task illuminance.

Local lighting can be a very efficient method for providing adequate task illumination, particularly where high illuminances are necessary and/or flexible directional lighting is required. Local lighting is frequently provided by luminaires mounted on the work station (e.g. desk lights).

Local lighting must be positioned to minimise shadows, veiling reflections and glare. Although local luminaires allow efficient utilisation of emitted light, the luminaires themselves may be inefficient and can be expensive. Most local lighting systems are accessible and often adjustable. This increases wear and tear and hence maintenance costs but provides some individual control.

Both local and localised lighting offer scope for switch control of individual luminaires which can be off when not required, but care must be taken with localised luminaires to ensure that sufficient ambient illumination is provided.

## 4.4.3  Choice of lamp and luminaire

The choice of lamp will affect the range of luminaires available, and vice-versa. Therefore, one cannot be considered without reference to the other. When planning an electric lighting scheme, designers try to select a single optimum lamp and luminaire combination which will meet their objectives. In experienced hands, or where there is a limited range of equipment available, this type of approach can be effective, but many designers remain uncertain as to the value of their solution and whether or not there is a better choice. One method of design which avoids this dilemma and draws attention to areas of difficulty, is to follow a procedure which does not try to identify a single lamp and luminaire combination but rather rejects those combinations which are unsatisfactory. In this manner, whatever remains will be acceptable and a final choice can be made by comparison. With such an approach unrealistic requirements will manifest themselves by causing all available choices to be eliminated. The need to juggle with different requirements in order to achieved a satisfactory compromise is avoided, since this will occur automatically. Finally, because all of the unrejected luminaire and lamp combinations are acceptable, the most efficient and economically acceptable scheme can be selected. This approach lends itself to computed aided design.

### 4.4.3.1 Choice of lamp

The designer should compile a list of suitable lamps, by rejecting those which do not satisfy the design objectives. The availability of suitable luminaires can then be checked and the economics of each assessed. General guidance can be obtained from section 3.2 of this Code.

The run-up time of most discharge lamps (excluding fluorescent lamps) is unsatisfactory for applications requiring rapid provision of illumination or switching unless auxiliary tungsten lamps are provided.

Lamps must have colour rendering properties suited to their intended use. Good colour rendering may be required in order to achieve better discrimination between colours where that is part of the visual task. Alternatively, good colour rendering may be required to achieve a particular appearance or degree of comfort (e.g. merchandising, offices, etc.).

A warm apparent colour tends to be preferred for informal situations, at lower illuminances and in cold environments, whilst a cool apparent colour tends to be preferred for formal situations, at higher illuminances, and in hot environments. Adjacent areas should not be lit with sources of significantly different apparent colour unless a special effect is required.

The life and lumen maintenance characteristics of the lamps must be considered to arrive at a practicable and economic maintenance schedule.

Where moving machinery is used care should be taken to avoid stroboscopic effects. All lamps operating on an alternating current exhibit some degree of cyclic variation of light output. It is most significant with discharge lamps which do not employ a phosphor coating. The problem can normally be reduced or eliminated by having alternate rows of luminaires on different phases of the supply and ensuring that critical areas receive illumination in roughly equal proportion from each phase. Alternatively the lamps may be operated from high frequency supplies, or illumination from local luminaires (with acceptable lamps that do not cause stroboscopic problems) can be used to swamp the general illumination.

When selecting a range of suitable lamps, the designer must consider the types of luminaires which are available and the degree of light control and light output required. Accurate light control is more difficult with large sources than with compact sources, however the latter will have a higher luminance (for the same output) and are potentially more glaring.

Standardisation of lamp types and sizes within a particular site or company can simplify maintenance and stocking.

### 4.4.3.2 Choice of luminaire

In the choice of luminaire, the designer can exercise a combination of professional judgement, personal preference and economic analysis. Luminaires may have to withstand a variety of physical conditions, such as vibration, moisture, dust, ambient temperature, vandalism and so on. In addition, the onus is on the designer to specify safe equipment. General guidance on the characteristics of luminaires can be obtained from section 3.3 of this Code.

Safety can be guaranteed by using equipment with the British Standard safety mark or obtaining written assurances from the manufacturer. It is important to ensure that equipment is selected which can withstand and operate safely in the environmental conditions that will be encountered. The Ingress Protection (IP) rating, which is discussed in section 3.3.1 gives guidance regarding the ability of the luminaire to withstand the ingress of solid foreign bodies and moisture. The designer must make sure that the manufacturer's claims apply throughout the intended life of the luminaire. This is particularly true of a claimed IP rating, which, without further qualification, applies to a new luminaire. However, when a luminaire has the British Standard safety mark, the IP rating is applied to an already arduously tested luminaire.

Not only must the luminaire withstand the ambient conditions, it may have to operate in a hazardous area, such as a refinery, mine or similar environment. In this event, special equipment is required to satisfy the safety regulations. Such equipment is beyond the scope of this Code. This subject is covered by the CIBS Application Guide: Lighting for hostile and hazardous environments.

The light distribution of the luminaire should be carefully considered as it influences the distribution of illuminance and the directional effects that will be achieved. To establish the nature of the distribution of illuminance and directional effects that will be achieved for a regular array of a given luminaire, the illuminance ratio (IR) charts of CIBS (IES) Technical Report No. 15, the Multiple Criterion Design Method, can be used. The use of these charts is described in section 4.4.3.3.

The utilisation factor (UF) for a luminaire is a measure of the efficiency with which light from the lamp is used for illuminating the working plane. The product of utilisation factor and lamp circuit luminous efficacy is the installed efficacy of the installation. In other words, it is a measure of how much luminous flux reaches the working plane for each watt of power applied. Luminaires can be ranked in order of the installed efficacy they provide, so that the most efficient luminaire, capable of meeting the other requirements, is selected. However, the luminaire with the highest installed efficacy may not offer the highest operating efficacy (see section 4.5.1.3). If a greater degree of switch control can be achieved with one type of lamp than another (e.g. tubular fluorescent versus high pressure sodium discharge), then the order may be reversed. This is because the hours of use may be sufficiently reduced in one case to offset its slightly lower installed efficacy. The ratio of the two installed efficacies is equal to the ratio of the hours of use which must be achieved for the schemes to have equal operating efficacies. Thus, if one scheme has twice the installed efficacy of a second scheme, then for the second scheme to have a better operating efficacy its hours of use must be 50% of that for the first scheme.

Luminaire reliability and life will have a direct impact on the economics of the scheme, and must be realistically considered. The ease with which luminaires can be installed and maintained will also affect the overall economics and convenience of the scheme. Luminaires with good maintenance characteristics and which can be easily maintained will not only save on maintenance costs, but will also be more efficient throughout their life.

Luminaires which can be unplugged and detached, or which have removable gear, can simplify maintenance by allowing remote servicing.

## 4.4.3.3 Illuminance Ratio (IR) Charts

Illuminance Ratio (IR) Charts are published in CIBS (IES) Technical Report No. 15, the Multiple Criterion Design Method. They enable the designer to examine the effects of room index, surface reflectances, luminaire direct ratio and flux fraction ratio upon illuminance ratios and the directional aspects of lighting.

They are published in pairs for different combinations of ceiling, wall and floor reflectance and for different room indices. The reflectances are given the symbols L, M and D to signify, light, medium and dark reflectances respectively. These correspond to:

|  | L | M | D |
|---|---|---|---|
| Ceiling cavity | 0.70 | 0.50 | 0.30 |
| Walls | 0.50 | 0.30 | 0.10 |
| Floor cavity | 0.30 | 0.20 | 0.10 |

Figure 4.7 shows a typical pair of charts. The charts are identical except for the loci plotted onto them. In each case the horizontal axis represents the direct ratio of the installation expressed as a British Zonal (BZ) number, and the vertical axis is the flux fraction ratio of the installation. Real luminaires can therefore be plotted onto the charts according to their flux fraction ratio and BZ number (N.B. BZ number may change with room index).

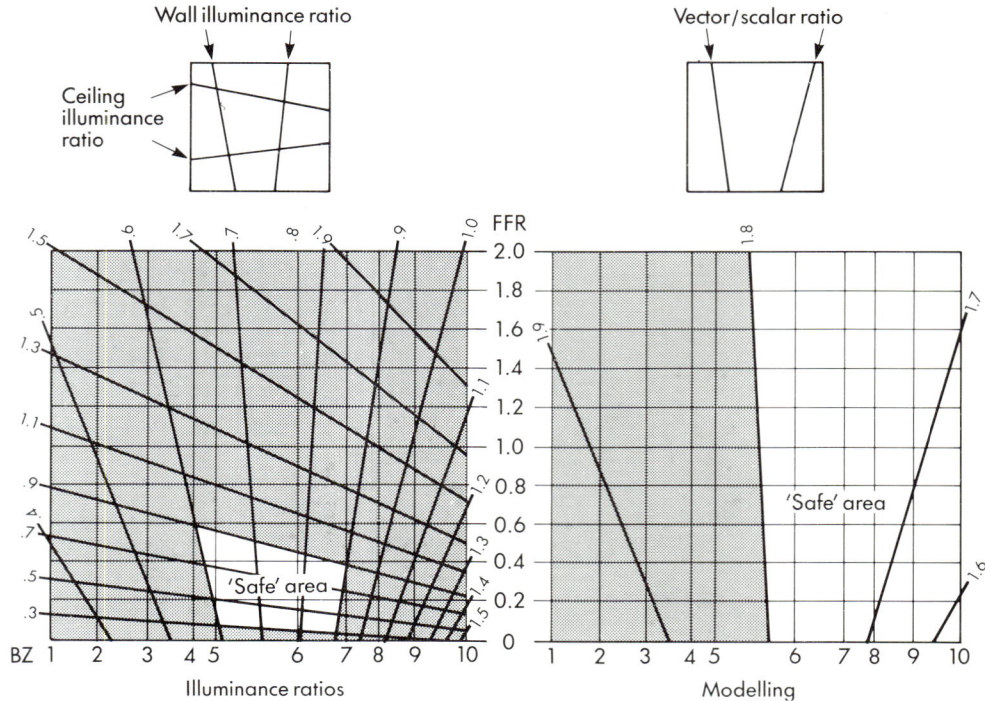

**Fig 4.7** A typical illuminance ratio (IR) chart

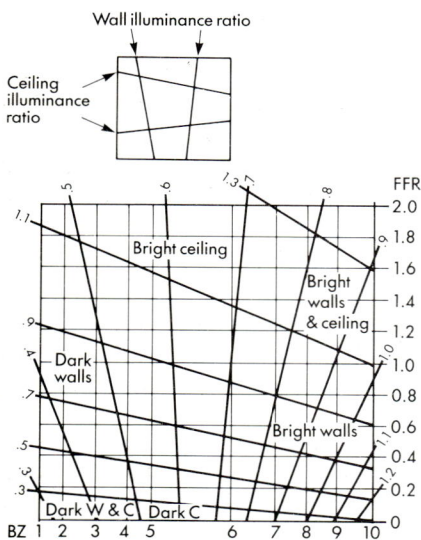

**Fig 4.8** The effect of illuminance ratios upon room appearance

Loci of constant illuminance ratio are plotted on the lefthand chart and loci of constant average vector/scalar ratio are plotted on the righthand chart.

The charts may be used in two ways. Luminaires can be plotted onto the charts to determine the illuminance ratios and average vector/scalar ratios that will be achieved by a regular array of such luminaires. Alternatively and more usefully at the general planning stage, the charts may be used to identify the range of reflectances and luminaires which can achieve the desired conditions. The range of acceptable luminaires can be identified by BZ numbers and flux fraction ratios. However, the process of selection will then be aided if the positions of real luminaires are marked onto the charts or a transparent overlay.

The recommended ranges of illuminance ratios and vector/scalar ratios are shown as unshaded (safe) areas on the charts. These recommendations are not sacrosanct and there are reasons why a designer may wish to deviate from them. Bright walls can make a room seem larger and more spacious. Dark walls can make it seem small and possibly cramped or intimate. Bright ceilings and dark walls may give the impression of formality and tension, whilst the reverse, bright walls and dark ceiling, may create an informal and relaxed or sociable atmosphere. These are not hard and fast rules, but are supported by experimentation. Figure 4.8 shows these tendencies mapped onto a typical IR Chart.

It is often impossible to simultaneously achieve the desired illuminance ratios and vector/scalar ratios without changes in reflectance, and even then it may still be impossible. Wall-washers may be employed to increase the wall to task illuminance ratio. The IR charts illustrate that for most cases a proportion of upward light from the luminaire is desirable to achieve

*General planning*

acceptable ceiling/task illuminance ratios.

### 4.4.4  System Management

A good lighting system must not only be well designed, but must also be managed and operated effectively and efficiently. System management must: *(a)* control the use of the system to ensure efficiency *(b)* maintain the system in good order.

### 4.4.4.1  Control

The lighting system must be designed and managed to permit good control of energy use. This is important during the working day and outside working hours (see section 3.4).

Methods of control fall into three broad categories: *(a)* Manual control (Managerial) *(b)* Automatic control (non intelligent) *(c)* Processor control (intelligent)

Manual methods rely upon individuals and appointed members of staff controlling the lighting system. These methods tend to be inexpensive in capital costs but may be less effective than automatic methods. To be effective the lighting system must be well planned to permit flexible switching of individual luminaires or banks of luminaires. The switch panels must be sensibly located and clearly marked (a mimic diagram can be very helpful). An education programme to ensure staff awareness is essential and this can be reinforced with posters, and with labels on or adjacent to the switch panels.

One of the main snags with manual methods is that, whilst occupants may be aware that natural lighting is insufficient and will turn on lights, it is rare for them to respond to sufficient daylight by turning lights off.

Automatic control in the form of an imposed switch-off (particularly at lunchtime) can be effective, since, if natural lighting is adequate, the luminaires may not be turned back on.

A considerable amount of energy is often wasted after working hours when the lights are left on to no useful purpose. The full lighting system may be on when cleaners are in the building. The provision of automatic cleaners' circuits controlling only some of the lighting to provide reduced illuminances can save money.

Automatic control systems, such as time switches or photocells, can be inexpensive and can switch (or dim) banks of lights. Photocells can monitor the level of useful daylight and turn off luminaires or individual lamps in rows adjacent to the windows. Whether or not this is economic will depend upon the daylight factor and the proportion of the working year for which the required illuminance is exceeded.

Time switches provide a convenient method of ensuring that unwanted lighting is not provided outside working hours.

Occupancy detectors can be used to detect the presence of occupants and to control the lights accordingly. These can rely upon acoustic, infra red, radar, or other methods of detection. A time-lag must normally be built into the system to prevent premature switch-offs.

Automatic systems must normally have some degree of manual override (on and off) to cater for unexpected circumstances. Systems which automatically cancel lighting but must be manually reset can offer greater savings than those which switch on again automatically. Occupants can always be relied upon to turn-on lighting if they need it.

Computer-based or microprocessor-based control systems are becoming increasingly popular, more reliable and less expensive. These rely upon dedicated computers or processors to control some or all of the building services. Lifts, fire alarms, lighting, air-conditioning and other equipment can be controlled. The most important advantages of such an approach are that complex decisions can be taken from moment to moment, based upon the precise state of the building's operation, and that the system is controlled by software. This last feature means that the control programs can be refined and tailored to suit the building and can be easily amended to suit changed circumstances.

Such intelligent systems can continuously monitor the building to operate it at maximum efficiency and economy. For example, lighting load can be shed in non-critical areas if the electricity maximum demand is reached during winter months, or shed in summer if cooling demands become excessive.

With any control system considerable care must be taken to ensure that acceptable lighting conditions are always provided for the occupants. Safety must always be of paramount importance.

Control systems which are obtrusive or disruptive are counter-productive and may even be sabotaged by the staff. For this reason, dimmer systems are often preferred. Photocells and other sensing circuitry must incorporate a delay to prevent sporadic and disruptive switch-offs, but must respond immediately when a switch-on is called for.

## 4.4.4.2 Maintenance

Lighting systems must be serviced regularly and this must be allowed for at the design stage. Faulty or failed lamps should be replaced and unsafe or faulty equipment should be rectified. In addition to this depreciation in illumination, caused by dirt on lamps, luminaires and room surfaces should be controlled by regular cleaning.

It is not always enough to replace lamps on failure. The light output of lamps falls with hours of operation. For most lamp types, a point will usually be reached where it is financially better to replace the lamps than to continue to waste power. Furthermore, labour charges for the replacement of individual lamps can be high, so it is often less expensive in the end to clean and service a complete installation when convenient, than to indulge in intermittent cleaning and relamping.

The effect of alterntive maintenance schedules on the illuminance provided by the installation can be examined by the procedure described in Appendix 7.

# 4.5 Detailed planning

When the overall design has been resolved in general terms, detailed calculations are required to determine such things as the number of luminaires, the Glare Index, the final cost and so on. Design calculations can be complex and the use of computers is widespread. It is easy in these circumstances to lose sight of the original objectives and purpose of the design. When the designer feels that he has completed the design he should stop and analyse how well the original objectives have been met.

He may, at this stage, find that the resultant design is unsatisfactory in some regard. This is by no means uncommon and reflects the inadequacy of the methods of design that are currently available. The only course of action is to revise the design until a suitable solution is found. This iterative procedure is a normal part of the design process.

The main calculations which may have to be carried out during the design process are detailed in the following sections.

## 4.5.1 Costs and energy use

The most powerful constraints on any design are financial, namely, how much will the scheme cost to install and operate.

Initially it is necessary to establish realistic economic and energy budgets commensurate with the design objectives. At all stages of the design, capital costs and running costs must be scrutinised and controlled. The economics and energy use of the lighting system must be considered as part of the building as a whole.

## 4.5.1.1 Financial evaluation

The methods of financial assessment employed by the designer must be acceptable to the client. This can cause difficulties, because grants, tax benefits, tariffs, accounting methods and other factors can vary. Fortunately simple methods of analysis are usually sufficient.

The designer must satisfy himself and others that: *(a)* a new scheme is justified, *(b)* that the proposed scheme is a sound economic proposition.

A new scheme may be justified because the building is new or because the existing scheme is no longer acceptable for its intended purpose. Alternatively, it may be necessary to decide whether it is better to retain existing equipment or replace it with a new scheme.

Scheme economics are difficult to judge in absolute terms. For this reason comparisons are often used. These can be against an existing scheme or an alternative design. If a comparison between alternative designs is to be meaningful, the schemes must be of equitable standard.

The cost of owning and operating an installation can be conveniently divided as follows:

*(a) Capital costs:* Lamps, luminaires and associated equipment, installation and wiring.

*(b) Operating costs:* (i) Fixed annual costs: Electricity supply charges (e.g. service charge, maximum demand charge, etc.) (ii) Running costs: electricity consumed (kWh), replacement lamps, relamping (labour). (iii) Maintenance costs: cleaning luminaires/lamp, repairing equipment

Many methods of scheme comparison are possible. The principles of several methods of financial evaluation are discussed in CIBS Guide, Section B18, Owning and Operating costs.

## 4.5.1.2 Energy and tariffs

A number of electricity tariffs are in use throughout the UK. Each Area Electricity Board has its own tariffs for different classes of consumer. The most common commercial and industrial payment systems are two-part tariffs and maximum demand tariffs. The two-part tariff usually consists of a rate per kWh of electricity supplied and a standing charge, based in some way on the maximum power required.

Typically, a maximum demand tariff has a maximum demand charge levied for each kVA (or kW with a power factor adjustment) of annual or monthly maximum connected load. A service capacity charge for each kVA or kW of installed load may also be levied. In addition to charges for the size of the connected load, a charge is made for each kWh of electricity consumed. This can be at a flat rate, or at separate day and night rates.

For both tariff types a fuel cost adjustment factor is usually incorporated. The Area Electricity Boards will offer advice on the most appropriate tariff for specific applications.

Control of the lighting load profile by switching or dimming, so that unnecessary lighting is not used, will reduce the amount of electricity consumed. In addition to this, maximum demand charges can be reduced by shedding lighting load at times of peak demand. Such peak demand often occurs during the middle of the day, when adequate daylight is available.

## 4.5.1.3 Energy calculations

All designers should ensure that their designs do not waste energy. However, the most important considerations about energy consumption are usually financial ones. Few users are willing to invest extra money to achieve energy savings unless the savings offer a reasonable rate of return on that investment. Functional lighting tends to be more efficient than decorative lighting. This is not to say that decorative lighting is wasteful. If the design objectives call for particular conditions to be created then they should be provided. If they are not provided, then, although the design may use less energy it will not be effective and cannot, therefore, be regarded as satisfactory.

It is easy to assume that the efficiency of an installation should be the most important yardstick in any design, and to give it undue bias. If this happens, not only will important design objectives be suppressed, but the resultant scheme is unlikely to be the best financial proposition. Scheme efficiency should therefore be considered in parallel with other design details.

**(1) Installed efficacy, operating efficacy and load factor**

A simple measure of the scheme efficiency is the installed efficacy. This is the product of the lamp circuit luminous efficacy and utilization factor.

Section 2.5 of this Code recommends ranges of installed efficacies appropriate for various applications.

It should be noted that where adequate daylight is available for part of the day, or where unwanted lighting can be switched off, the operating efficacy of the scheme can be higher than the installed efficacy. The ratio of the installed efficacy to the operating efficacy is the load factor. The load factor for a lighting installation, during a specified period of time, is the ratio of the energy actually consumed to the energy that would have been consumed had this occurred at the maximum rate throughout the specified period. Thus if 25% of the lights in an installation are switched off on average throughout the working day, the load factor will be 0.75. Hence the operating efficacy will be 1.33 times the installed efficacy.

For many installations the load factor will be determined by the ability of the lighting control system to switch the lighting in response to daylight availability. To compare the effectiveness of alternative control systems, a designer will need to estimate the probable annual use of electric lighting under each system. The notes below deal separately with traditional switching arrangements – a panel of switches by the entrance, and energy saving alternatives.

**(2) Conventional switching**

Field studies of occupants' switching behaviour have shown that, with traditional switching arrangements, electric lighting is usually either all switched-on or all switched-off. The act of switching is almost entirely confined to the beginning and end of a period of occupation; people may switch lights on when entering a room but seldom turn them off until they all leave. The year-round probability that an occupant will switch lights on when entering a room depends on the time of day, the orientation of the window and the minimum orientation-weighted daylight factor on the working area, (Fig 4.9). (See Appendix 9 for daylight factor calculations.) For example, if the minimum orientation-weighted daylight factor is 0.6 per cent and work starts at 9 a.m., figure 4.9 shows a 56 per cent probability of switching. If the room is in continual occupation, even through the lunch hour, we may conclude that this same figure – 56 per cent – is the probability that the lights will be on at any moment during the working day.

In rooms which empty for lunch, the lighting will probably be switched off by the last person out. Here it would be reasonable to treat the periods before and after lunch separately. In the example cited above, if lunch ends at 1.30 p.m., the probability of lights being on after lunch is taken as 37 per cent, during lunch as 0 per cent and during the morning as 56 per cent, as before.

If luminaires are logically zoned with respect to the natural lighting with pull-cord switches handy for the occupants to use, one can treat each zone as a separate room. The probability of switching would differ from zone to zone, depending on the minimum orientation weighted daylight factor in each zone. Figure 4.9 would still be applicable but the minimum orientation weighted daylight factor, and consequent energy savings, must be estimated separately for each zone.

A room occupied intermittently can be treated similarly but some assumption must be made about the periods when the space will be empty.

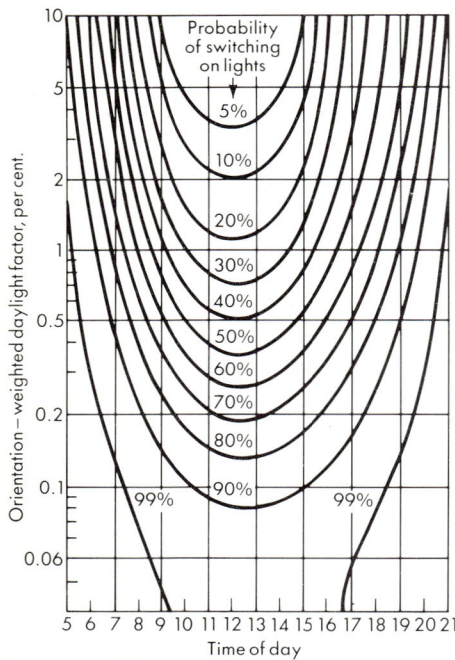

**Fig 4.9** The probability of switching luminaires on (after Hunt, D.R.G., L.R. and T., **12**(1), 7 (1980))

*Detailed planning*

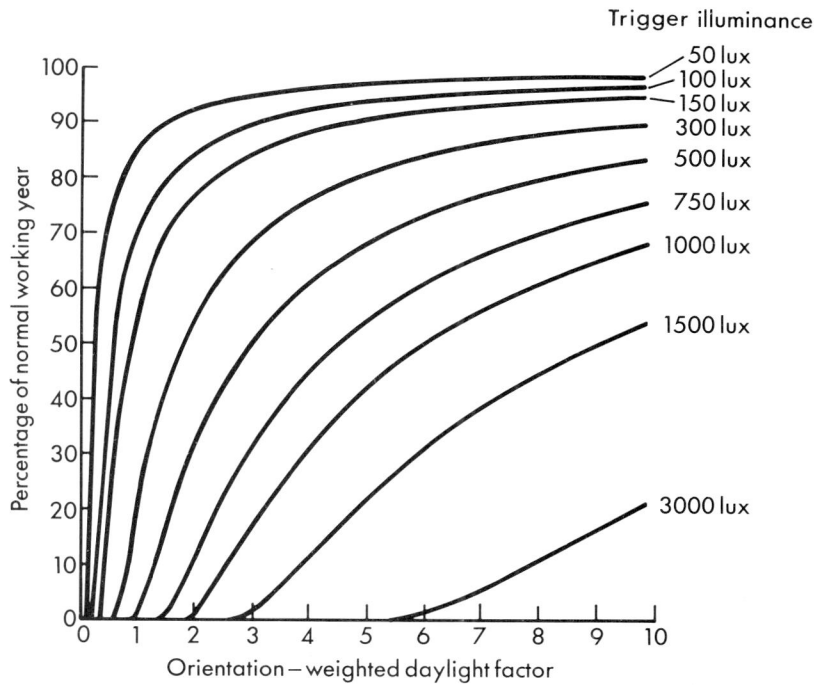

**Fig 4.10** Percentage of a normal working year during which luminaires will be switched off

(after Hunt, D.R.G., L.R. and T., **11**(1), 1 (1979))

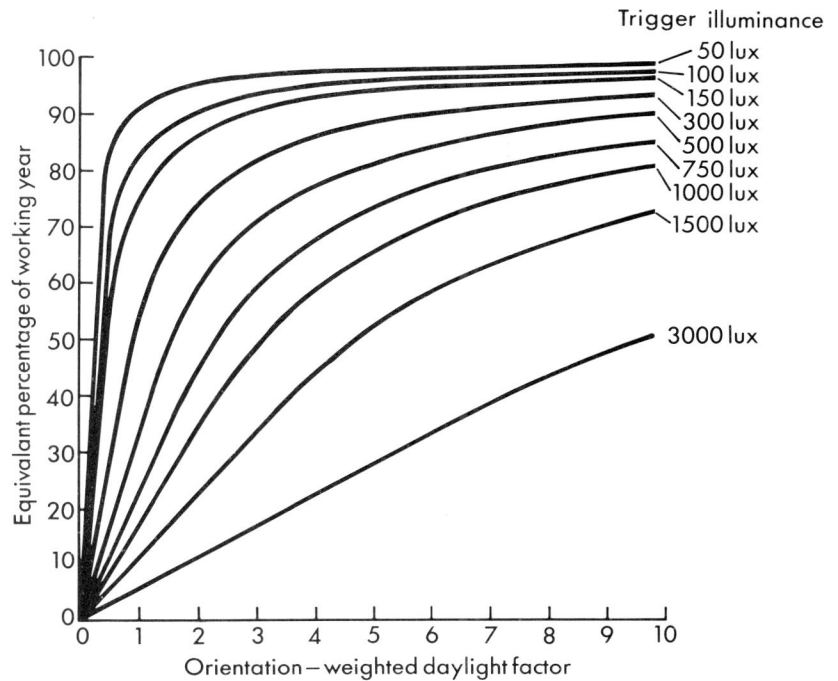

**Fig 4.11** Percentage of a normal working year during which luminaires would have to be switched off in order to ensure the same energy saving as dimming (top-up control)

(after Hunt, D.R.G., L.R. and T., **11**(1), 1 (1979))

## (3) Photo-electric on/off switching control

Photo-electric controls will normally be zoned to take full advantage of daylight. Figure 4.10 shows the percentage of a normal working year during which the luminaires would be off, as a function of the orientation-weighted daylight factor and of the illuminance at which the luminaires are switched, i.e. the 'trigger' illuminance. These curves assume that 'on' and 'off' switching will occur at the same illuminance. Where this is not the case, i.e. where the luminaires are switched off at an illuminance appreciably greater than that at which they are switched on, the mean of the two illuminances should be taken as the 'trigger' illuminance for Fig 4.10.

**(4) Photo-electric dimming ('top-up'/control)**

Estimation of energy saving from continuous dimming is complicated by the fact that the lamp circuit luminous efficacy generally decreases as a lamp is dimmed. For a well-designed tubular fluorescent dimming circuit, the cathode heaters consume some 12% of the nominal power consumption and the remaining wattage is roughly proportional to the light output.

Figure 4.11 has been constructed on this basis. It shows the percentage of a normal working year during which the luminaires would have to be switched right off in order to ensure the energy saving obtainable by continuous photo-electric dimming.

### 4.5.2 Light loss caused by dirt and depreciation

The initial performance of an installation when new will not be realised throughout the life of the installation because the components of the installation deteriorate to varying degres. The following are important factors. *(a)* Lamp lumen depreciation, i.e. the light output of lamps will fall with hours of use. *(b)* Luminaire dirt depreciation, i.e., the build-up of dirt on the luminaire and lamp will reduce the light output. This will depend upon the nature of the luminaire construction, the amount and type of airborne pollution and the cleaning interval. *(c)* Room surface dirt depreciation, i.e. the accumulation of dirt on room surfaces will change the amount of inter-reflected light.

The total effect of these factors for different situations can be consolidated into a single 'light loss factor' by the method described in appendix 7.

### 4.5.3 Average illuminance

The average illuminance produced by a lighting installation, or the number of luminaires required to achieve the average illuminance, can be calculated by means of utilisation factors.

### 4.5.3.1 Utilisation factors

The utilisation factor *UF(S)* of an installation is the ratio of the total flux received by the reference surface *S* to the total lamp flux of the installation.

The average illuminance *E(S)* over the reference surface *S* can therefore be calculated from the 'lumen method' formula:

$$E(S) = \frac{F \times n \times N \times LLF \times UF(S)}{A(S)}$$

where   $F$ = the initial bare lamp luminous flux (lumens)
$n$ = the number of lamps per luminaire
$N$ = the number of luminaires
$LLF$ = the light loss factor (see appendix 7)
$UF(S)$ = the utilisation factor for the reference surface S
$A(S)$ = the area of the reference surface S (metre$^2$)

The formula can be re-arranged to permit the calculation of the number of luminaires required to achieve a chosen illuminance.

Utilisation factors can be determined for any surface or layout of luminaires, but, in practice, are only calculated for general lighting systems with regular arrays of luminaires and for the three main room surfaces, the ceiling cavity, the walls, and the floor cavity (see Fig 4.12). Utilisation factors for these surfaces: the ceiling cavity, the walls and the floor cavity or horizontal reference plane, are designated *UF(C)*, *UF(W)* and *UF(F)* respectively. The method for calculating utilisation factor for these surfaces is given in CIBS Technical Memorandum 5: The calculation and use of utilisation factors.

Although utilisation factors can be calculated by the lighting designer, most manufacturers publish utilisation factors for standard conditions for their luminaires. CIBS Technical Memorandum No. 5 defines a standard method of presentation and states the assumptions on which the tabulated values are based. Figure 4.13 is an example of the standard presentation.

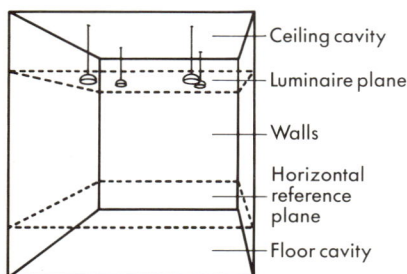

**Fig 4.12** Ceiling cavity, walls and floor cavity

- Ceiling cavity
- Luminaire plane
- Walls
- Horizontal reference plane
- Floor cavity

| Utilisation Factors UF(F) Standard Presentation | | | | | | | | | | | SHR NOM = 1.5 | |
|---|---|---|---|---|---|---|---|---|---|---|---|---|
| Room Reflectance | | | Room Index | | | | | | | | | |
| C | W | F | 0.75 | 1.00 | 1.25 | 1.50 | 2.00 | 2.50 | 3.00 | 4.00 | 5.00 | |
| 0.70 | 0.50 | 0.20 | 0.43 | 0.49 | 0.55 | 0.60 | 0.66 | 0.71 | 0.75 | 0.80 | 0.83 | |
| | 0.30 | | 0.35 | 0.41 | 0.47 | 0.52 | 0.59 | 0.65 | 0.69 | 0.75 | 0.78 | |
| | 0.10 | | 0.29 | 0.35 | 0.41 | 0.46 | 0.53 | 0.59 | 0.63 | 0.70 | 0.74 | |
| 0.50 | 0.50 | 0.20 | 0.38 | 0.44 | 0.49 | 0.53 | 0.59 | 0.63 | 0.66 | 0.70 | 0.73 | |
| | 0.30 | | 0.31 | 0.37 | 0.42 | 0.46 | 0.53 | 0.58 | 0.61 | 0.66 | 0.70 | |
| | 0.10 | | 0.27 | 0.32 | 0.37 | 0.41 | 0.48 | 0.53 | 0.57 | 0.62 | 0.66 | |
| 0.30 | 0.50 | 0.20 | 0.30 | 0.37 | 0.41 | 0.45 | 0.52 | 0.57 | 0.60 | 0.65 | 0.69 | |
| | 0.30 | | 0.28 | 0.33 | 0.38 | 0.41 | 0.47 | 0.51 | 0.54 | 0.59 | 0.62 | |
| | 0.10 | | 0.24 | 0.29 | 0.34 | 0.37 | 0.43 | 0.48 | 0.51 | 0.56 | 0.59 | |
| 0.00 | 0.00 | 0.00 | 0.19 | 0.23 | 0.27 | 0.30 | 0.35 | 0.39 | 0.42 | 0.46 | 0.48 | |

Rating: 65W 1500 mm  Mounted: On Ceiling
Multiply UF values by service correction factors
Calculated in accordance with CIBS Technical Memorandum No. 5/1980

**Fig 4.13**  Typical utilisation factor (UF(F)) table for a twin fluorescent fitting

## 4.5.3.2  Defining the main room surfaces

To use utilisation factor tables it is necessary to have values of the Room Index and the room reflectances. The room is considered to consist of three main surfaces: the ceiling cavity, the walls, and the floor cavity or horizontal reference plane (Fig 4.12).

## 4.5.3.3  Room index

The room index is a measure of the proportions of the room. It is twice the plan area of the room divided by the area of the walls between the horizontal reference plane (mouth of the floor cavity) and the luminaire plane (mouth of the ceiling cavity) – (Fig 4.12).
For rectangular rooms the room index $RI$ is:

$$RI = \frac{L\,W}{(L + W)\,H}$$

where  $L$ = the length of the room
$W$ = the width of the room
$H$ = the height of the luminaire plane above the horizontal reference plane

Results may be rounded to the nearest value in the utilisation factor table.
If the room is re-entrant in shape (e.g. L-shaped) then it must be divided into two or more non re-entrant sections which can be treated separately.

## 4.5.3.4  Effective reflectance

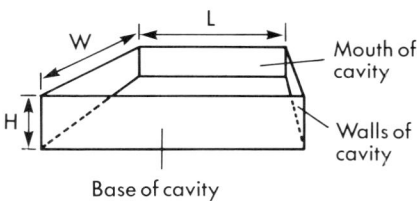

**Fig 4.14**  Calculation of cavity index

In order to use utilisation factor tables correctly the effective reflectances of the ceiling cavity, walls and floor cavity must be calculated.
For the ceiling cavity and floor cavity the cavity indices $CI(C)$ and $CI(F)$ must be calculated. The cavity index $(CI)$, which is similar in concept to the room index, is given by: (Fig 4.14)

$$CI = \frac{\text{mouth area of cavity} + \text{base area of cavity}}{\text{wall area of cavity}}$$

For rectangular rooms
$$CI(C) \text{ or } CI(F) = \frac{L\,W}{(L + W)\,h} = RI.\,H/h$$

Where $h$ is the depth of the cavity.
The effective reflectance $RE(X)$ of the cavity $X$ can then be determined from Fig 4.15 or from the simplified, but less accurate, formula:

$$RE(X) = \frac{CI(X) \times RA(X)}{CI(X) + 2\,(1 - RA(X))}$$

where  $RA(X)$ is the average area weighted reflectance of the cavity $X$
$CI(X)$ is the cavity index of the cavity $X$

| Reflectance of Cavity | | Cavity Index | | | | | | | | | |
|---|---|---|---|---|---|---|---|---|---|---|---|
| Walls | Base | 1 | 2 | 3 | 4 | 5 | 6 | 7 | 8 | 9 | 10 |
| 0.1 | 0.1 | 0.037 | 0.056 | 0.067 | 0.073 | 0.078 | 0.081 | 0.083 | 0.085 | 0.087 | 0.088 |
| | 0.2 | 0.056 | 0.098 | 0.122 | 0.137 | 0.148 | 0.155 | 0.161 | 0.165 | 0.168 | 0.171 |
| | 0.3 | 0.075 | 0.140 | 0.178 | 0.201 | 0.218 | 0.229 | 0.238 | 0.245 | 0.250 | 0.255 |
| | 0.4 | 0.094 | 0.182 | 0.234 | 0.266 | 0.288 | 0.303 | 0.315 | 0.325 | 0.332 | 0.338 |
| | 0.5 | 0.113 | 0.224 | 0.289 | 0.330 | 0.358 | 0.378 | 0.393 | 0.404 | 0.414 | 0.422 |
| | 0.6 | 0.132 | 0.267 | 0.345 | 0.395 | 0.428 | 0.452 | 0.470 | 0.484 | 0.496 | 0.505 |
| | 0.7 | 0.151 | 0.309 | 0.401 | 0.459 | 0.498 | 0.527 | 0.548 | 0.565 | 0.578 | 0.589 |
| | 0.8 | 0.171 | 0.352 | 0.458 | 0.524 | 0.569 | 0.601 | 0.626 | 0.645 | 0.660 | 0.673 |
| 0.2 | 0.1 | 0.058 | 0.073 | 0.080 | 0.084 | 0.087 | 0.089 | 0.090 | 0.092 | 0.092 | 0.093 |
| | 0.2 | 0.079 | 0.117 | 0.137 | 0.150 | 0.158 | 0.164 | 0.169 | 0.172 | 0.175 | 0.177 |
| | 0.3 | 0.100 | 0.161 | 0.195 | 0.216 | 0.230 | 0.240 | 0.247 | 0.253 | 0.258 | 0.262 |
| | 0.4 | 0.121 | 0.206 | 0.253 | 0.282 | 0.301 | 0.316 | 0.326 | 0.334 | 0.341 | 0.346 |
| | 0.5 | 0.142 | 0.250 | 0.311 | 0.348 | 0.373 | 0.391 | 0.405 | 0.416 | 0.424 | 0.431 |
| | 0.6 | 0.163 | 0.296 | 0.369 | 0.415 | 0.446 | 0.468 | 0.484 | 0.497 | 0.507 | 0.516 |
| | 0.7 | 0.185 | 0.341 | 0.428 | 0.482 | 0.518 | 0.544 | 0.563 | 0.579 | 0.591 | 0.601 |
| | 0.8 | 0.207 | 0.386 | 0.487 | 0.549 | 0.591 | 0.620 | 0.643 | 0.660 | 0.674 | 0.686 |
| 0.3 | 0.1 | 0.082 | 0.091 | 0.094 | 0.095 | 0.096 | 0.097 | 0.098 | 0.098 | 0.098 | 0.098 |
| | 0.2 | 0.105 | 0.137 | 0.153 | 0.163 | 0.169 | 0.174 | 0.177 | 0.180 | 0.182 | 0.184 |
| | 0.3 | 0.128 | 0.184 | 0.213 | 0.231 | 0.242 | 0.251 | 0.257 | 0.262 | 0.266 | 0.269 |
| | 0.4 | 0.151 | 0.231 | 0.273 | 0.299 | 0.316 | 0.328 | 0.337 | 0.344 | 0.350 | 0.355 |
| | 0.5 | 0.175 | 0.278 | 0.334 | 0.367 | 0.390 | 0.406 | 0.418 | 0.427 | 0.434 | 0.440 |
| | 0.6 | 0.199 | 0.326 | 0.395 | 0.436 | 0.464 | 0.484 | 0.498 | 0.510 | 0.519 | 0.526 |
| | 0.7 | 0.223 | 0.375 | 0.456 | 0.506 | 0.539 | 0.562 | 0.579 | 0.593 | 0.604 | 0.613 |
| | 0.8 | 0.248 | 0.424 | 0.518 | 0.575 | 0.613 | 0.641 | 0.661 | 0.676 | 0.689 | 0.699 |
| 0.4 | 0.1 | 0.107 | 0.109 | 0.108 | 0.107 | 0.106 | 0.105 | 0.105 | 0.104 | 0.104 | 0.104 |
| | 0.2 | 0.133 | 0.158 | 0.170 | 0.176 | 0.181 | 0.184 | 0.186 | 0.187 | 0.189 | 0.190 |
| | 0.3 | 0.159 | 0.208 | 0.232 | 0.246 | 0.255 | 0.262 | 0.267 | 0.271 | 0.274 | 0.276 |
| | 0.4 | 0.185 | 0.258 | 0.295 | 0.316 | 0.331 | 0.341 | 0.349 | 0.354 | 0.359 | 0.363 |
| | 0.5 | 0.211 | 0.308 | 0.358 | 0.387 | 0.407 | 0.420 | 0.431 | 0.439 | 0.445 | 0.450 |
| | 0.6 | 0.239 | 0.360 | 0.422 | 0.459 | 0.483 | 0.500 | 0.513 | 0.523 | 0.531 | 0.537 |
| | 0.7 | 0.266 | 0.412 | 0.486 | 0.531 | 0.560 | 0.581 | 0.596 | 0.608 | 0.617 | 0.625 |
| | 0.8 | 0.294 | 0.464 | 0.552 | 0.603 | 0.637 | 0.661 | 0.679 | 0.693 | 0.704 | 0.713 |
| 0.5 | 0.1 | 0.136 | 0.129 | 0.123 | 0.119 | 0.116 | 0.114 | 0.112 | 0.111 | 0.110 | 0.109 |
| | 0.2 | 0.164 | 0.181 | 0.187 | 0.190 | 0.192 | 0.194 | 0.195 | 0.195 | 0.196 | 0.196 |
| | 0.3 | 0.193 | 0.233 | 0.252 | 0.262 | 0.269 | 0.274 | 0.277 | 0.280 | 0.282 | 0.284 |
| | 0.4 | 0.223 | 0.287 | 0.317 | 0.335 | 0.346 | 0.354 | 0.360 | 0.365 | 0.369 | 0.372 |
| | 0.5 | 0.253 | 0.341 | 0.383 | 0.408 | 0.424 | 0.436 | 0.444 | 0.450 | 0.456 | 0.460 |
| | 0.6 | 0.284 | 0.396 | 0.450 | 0.482 | 0.503 | 0.517 | 0.528 | 0.537 | 0.543 | 0.548 |
| | 0.7 | 0.316 | 0.452 | 0.518 | 0.557 | 0.582 | 0.600 | 0.613 | 0.623 | 0.631 | 0.637 |
| | 0.8 | 0.348 | 0.509 | 0.587 | 0.633 | 0.662 | 0.683 | 0.698 | 0.710 | 0.719 | 0.727 |
| 0.6 | 0.1 | 0.168 | 0.151 | 0.139 | 0.131 | 0.126 | 0.123 | 0.120 | 0.118 | 0.116 | 0.115 |
| | 0.2 | 0.200 | 0.205 | 0.205 | 0.205 | 0.204 | 0.204 | 0.204 | 0.203 | 0.203 | 0.203 |
| | 0.3 | 0.232 | 0.261 | 0.273 | 0.279 | 0.283 | 0.286 | 0.288 | 0.289 | 0.290 | 0.291 |
| | 0.4 | 0.266 | 0.318 | 0.341 | 0.354 | 0.362 | 0.368 | 0.372 | 0.376 | 0.378 | 0.380 |
| | 0.5 | 0.301 | 0.376 | 0.410 | 0.430 | 0.443 | 0.451 | 0.458 | 0.463 | 0.467 | 0.470 |
| | 0.6 | 0.336 | 0.435 | 0.481 | 0.507 | 0.524 | 0.535 | 0.544 | 0.550 | 0.556 | 0.560 |
| | 0.7 | 0.373 | 0.496 | 0.552 | 0.585 | 0.605 | 0.620 | 0.630 | 0.639 | 0.645 | 0.650 |
| | 0.8 | 0.411 | 0.557 | 0.625 | 0.663 | 0.688 | 0.705 | 0.718 | 0.727 | 0.735 | 0.741 |
| 0.7 | 0.1 | 0.204 | 0.173 | 0.155 | 0.144 | 0.137 | 0.132 | 0.128 | 0.125 | 0.122 | 0.120 |
| | 0.2 | 0.240 | 0.231 | 0.224 | 0.220 | 0.217 | 0.215 | 0.213 | 0.211 | 0.210 | 0.210 |
| | 0.3 | 0.277 | 0.291 | 0.295 | 0.296 | 0.297 | 0.298 | 0.298 | 0.299 | 0.299 | 0.299 |
| | 0.4 | 0.315 | 0.352 | 0.366 | 0.374 | 0.379 | 0.382 | 0.385 | 0.387 | 0.388 | 0.389 |
| | 0.5 | 0.355 | 0.414 | 0.439 | 0.453 | 0.461 | 0.468 | 0.472 | 0.475 | 0.478 | 0.480 |
| | 0.6 | 0.397 | 0.478 | 0.513 | 0.533 | 0.545 | 0.554 | 0.560 | 0.565 | 0.568 | 0.571 |
| | 0.7 | 0.440 | 0.543 | 0.589 | 0.614 | 0.630 | 0.641 | 0.649 | 0.655 | 0.660 | 0.663 |
| | 0.8 | 0.485 | 0.611 | 0.666 | 0.696 | 0.715 | 0.728 | 0.738 | 0.745 | 0.751 | 0.756 |
| 0.8 | 0.1 | 0.245 | 0.198 | 0.173 | 0.158 | 0.148 | 0.141 | 0.136 | 0.132 | 0.129 | 0.126 |
| | 0.2 | 0.285 | 0.260 | 0.245 | 0.236 | 0.230 | 0.225 | 0.222 | 0.220 | 0.218 | 0.216 |
| | 0.3 | 0.328 | 0.323 | 0.318 | 0.315 | 0.312 | 0.311 | 0.309 | 0.308 | 0.308 | 0.307 |
| | 0.4 | 0.373 | 0.388 | 0.393 | 0.395 | 0.396 | 0.397 | 0.398 | 0.398 | 0.398 | 0.399 |
| | 0.5 | 0.419 | 0.456 | 0.469 | 0.477 | 0.481 | 0.484 | 0.487 | 0.488 | 0.490 | 0.491 |
| | 0.6 | 0.468 | 0.525 | 0.548 | 0.560 | 0.567 | 0.573 | 0.576 | 0.579 | 0.582 | 0.583 |
| | 0.7 | 0.519 | 0.596 | 0.627 | 0.644 | 0.655 | 0.662 | 0.667 | 0.671 | 0.674 | 0.677 |
| | 0.8 | 0.573 | 0.670 | 0.709 | 0.730 | 0.743 | 0.752 | 0.759 | 0.764 | 0.768 | 0.771 |

**Fig 4.15** The effective reflectance of cavities. To use this figure it is necessary to know the reflectance of the surfaces forming the cavity and the geometry of the cavity

The average reflectance RA(X) of a series of surfaces S1 to Sn with reflectances R(Sn) and areas A1 to An respectively is given by:

$$RA(X) = \frac{\sum\limits_{1}^{n} R(S_k)\, A(k)}{\sum\limits_{1}^{n} A(k)}$$

It should be noted that in order to calculate the effective reflectances, it is not necessary to know the colours of the surfaces, only the reflectances are required.

*Detailed planning*

## 4.5.3.5 Maximum spacing to height ratio

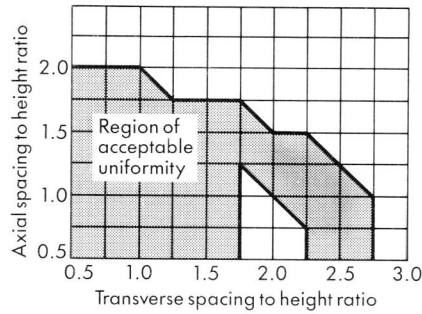

**Fig 4.16** Combination of transverse and axial spacing to height ratios producing acceptable uniformity (batwing luminaire)

The maximum spacing to height ratio *(SHR MAX)* of the luminaire is normally calculated when the utilisation factor table is calculated. The maximum spacing between the centres of luminaires divided by the mounting height above the horizontal reference plane should not exceed *SHR MAX* if uniformity of illuminance is to be acceptable for general lighting.

For some luminaires, notably those with distinctly di-symmetric distributions, extra spacing to height ratio information may be given.

The best form of information is a graph of acceptable *SHR* combinations in the axial and transverse directions. Figure 4.16 is an example of this for a particular luminaire.

For linear luminaires with conventional distributions, the maximum spacing to height ratio *SHR MAX* can be supplemented by the maximum transverse spacing to height ratio *(SHR MAX TR)*. This approach is less precise than the graphical method. The axial spacing to height ratio *(SHR AX)* should not exceed *SHR MAX* and the transverse spacing to height ratio *SHR TR* should not exceed the maximum transverse spacing to mounting height ratio *SHR MAX TR*. In addition to this, the product of *SHR AX* and *SHR TR* should not exceed *SHR MAX* squared. Thus:

$$SHR\ AX \times SHR\ TR \leqslant (SHR\ MAX)^2$$
$$\text{and} \quad SHR\ AX \leqslant SHR\ MAX$$
$$\text{and} \quad SHR\ TR \leqslant SHR\ MAX\ TR$$

## 4.5.3.6 Calculation procedure

The following procedure gives guidance on the sequence of calculations to be performed when calculating the number of luminaires necessary to obtain a chosen average illuminance on the horizontal reference plane by the lumen method.

*1*  Calculate the room index *RI*, the floor cavity index *CI(F)* and the ceiling cavity index *CI(C)*. (See section 4.5.3.3. and 4.5.3.4.)

*2*  Calculate the effective reflectances of the ceiling cavity, walls and floor cavity. Remember to include the effect of desks or machines in the latter. (See section 4.5.3.4.)

*3*  Determine the utilisation factor value from the manufacturer's data for the luminaire, using the room index and effective reflectances calculated as above.

*4*  Apply any correction factors, given in the utilisation factor table for lamp type or mounting position, to the utilisation factor *UF(F)* value.

*5*  Determine the light loss factor (see appendix 7).

*6*  Insert the appropriate variables into the lumen method formula:

$$N = \frac{E(F) \times A(F)}{F \times n \times LLF \times UF(F)}$$

to obtain the number of luminaires required.

where  *E(F)*  = the average illuminance to be provided on the working plane (lux)
  *A(F)*  = the area of the working plane (metre$^2$)
  *F*  = the initial bare lamp luminous flux (lumens)
  *n*  = the number of lamps/luminaire
  *LLF*  = the light loss factor
  *UF(F)*  = the utilisation factor for the plane
  *((F)* – refers to horizontal reference plane)

*7*  Determine a suitable layout.

*8*  Check that the geometric mean spacing to height ratio of the layout is within the range of the nominal spacing to height ratio *(SHR NOM)* for which the utilisation factor table is based, i.e.

$$\sqrt{(SHR\ AX \times SHR\ TR)} = SHR\ NOM \pm 0.5$$

9    Check that the proposed layout does not exceed the maximum spacing to height ratios. Namely:
Either – Check the value of *SHR AX* and *SHR TR* against a uniformity graph
or – Check that:

| | *SHR AX* × *SHR TR* | $\leqslant (SHR\ MAX)^2$ |
|---|---|---|
| and | *SHR AX* | $\leqslant SHR\ MAX$ |
| and | *SHR TR* | $\leqslant SHR\ MAX\ TR^*$ |

*or *SHR MAX* if *SHR MAX TR* not given.

10   Calculate the illuminance that will be achieved by the final layout.

### 4.5.4   Illuminance at a point

When local or localised lighting systems are employed or when unusual layouts or luminaires with unconventional light distributions are used, calculations of average illuminance can be inadequate or meaningless. In such circumstances it is necessary to calculate the illuminance at all points of interest.

These calculations can be done in one or more of three ways. *(a)* Calculations by hand from basic photometric data, *(b)* calculation from pre-calculated aids such as isolux diagrams, *(c)* calculation using a computer program.

The first method involves the greatest amount of work and is only suitable for a few points before it becomes tedious. However, hand calculations, when used with discretion, can yield much information about the required solution.

Isolux diagrams (Fig 4.17), if available for the particular set of circumstances, offer a faster method of carrying out the same calculations. For local and localised lighting systems they can provide considerable guidance on the correct location of luminaires. Isolux diagrams must normally be produced for the mounting height, scale and lumen output required. If this is not done, then, although isolux diagrams can still be used, considerable correction is necessary.

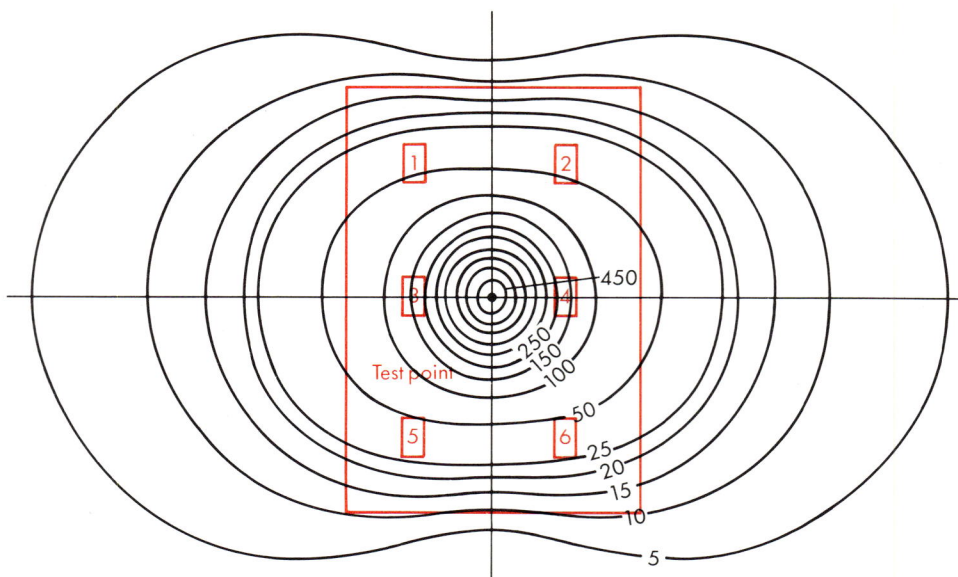

**Fig 4.17**  Use of Isolux Diagram to obtain total direct illuminance. By placing the central point of the Isolux Diagram over a given point on a plan of the room to the same scale, the illuminance at that point can be calculated.

In the example the contributions from luminaires 1 to 6 are 40, 40, 150, 160, 40 and 40 lux. Thus the total illuminance is 470 lux

Where a computer is available with suitable programs, illuminance values can be easily calculated. Although some limited design programs do exist, most programs simulate the illuminance pattern produced by a chosen layout of luminaires.

The ease with which computers can be used often results in abuse. The quality of the results is only as good as the calculation approach used and the data on which the calculation is performed, but the computer printout gives the impression of precision. Before using a computer program the designer should take care that the assumptions contained within the program are understood and should ensure that the data used is appropriate to the equipment and the situation of interest.

At the planning stage it is often better to obtain illuminance plots for individual luminaires or a group of luminaires than to attempt to simulate the complete installation. The performance of an individual luminaire or group can be easily analysed to assist with selection of the best layout. When the complete layout has been established it can be simulated on the computer.

The output data is important. If the information is insufficiently detailed, then important features may be overlooked. More commonly, if the data is too detailed it becomes too complex to interpret.

Graphic methods such as contour maps or boundary maps are easy to understand, and are preferable to tabulated data (Fig 4.18). However, they require more sophisticated software and hardware than conventional printouts.

**4.5.4.1 Illuminance at a point calculations**

The direct illuminance at a point can be calculated by the inverse square law. However, this only applies when the source is small compared with the distance between it and the point of illumination. When this is not the case, the calculation must be modified. Sources can be considered to be one of three basic types:

**4.5.4.2 Point sources**

A luminaire can be considered as a point source if its largest dimension is less than a fifth of the distance from it to the point being illuminated.

For a point source, the direct illuminance at a point can be found by applying the inverse square and cosine laws. See appendix 10 for formulae.

**4.5.4.3. Line sources**

When a luminaire is too long to be considered as a point source it can be regarded as a line source, providing that its width is less than a fifth of the distance from its centre to the point being illuminated. For most situations, fluorescent luminaires come into this category. To perform calculations on line sources it is necessary to have some method of integrating the effect of the length of the luminaire. One such method of doing this is the 'Aspect Factor Method'.

Aspect factors are derived from the axial luminous intensity distribution of the luminaire and, when used in the correct formula, make allowance for the effect of the length of the luminaire. Aspect factors can be published for almost all fluorescent luminaires, and must be used with the transverse luminous intensity distribution curves.

Two sets of aspect factors are provided. The parallel plane aspect factors are for calculating the direct illuminance on surfaces parallel to the axis of the luminaire, such as the floor, working plane or side walls. The other set of aspect factors, the perpendicular plane aspect factors are for calculating the direct illuminances on surfaces normal (i.e. at right angles) to the axis of the luminaire, such as the end walls (see appendix 10 for formulae).

Surfaces which are neither parallel to, nor perpendicular to, the axis of the luminaire (such as an angled drawing board) can be dealt with by a combination of the two types of aspect factor (see appendix 10).

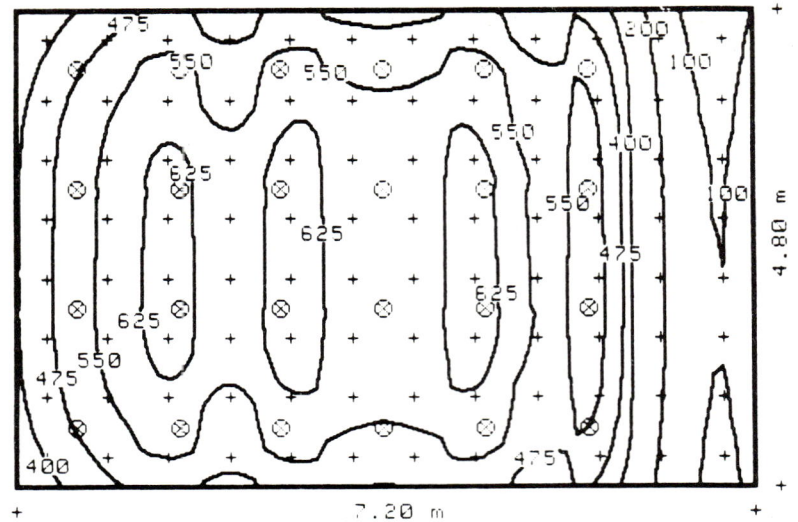

Fig 4.18 Contour maps and surfaces of illuminance in an interior as drawn by a computer

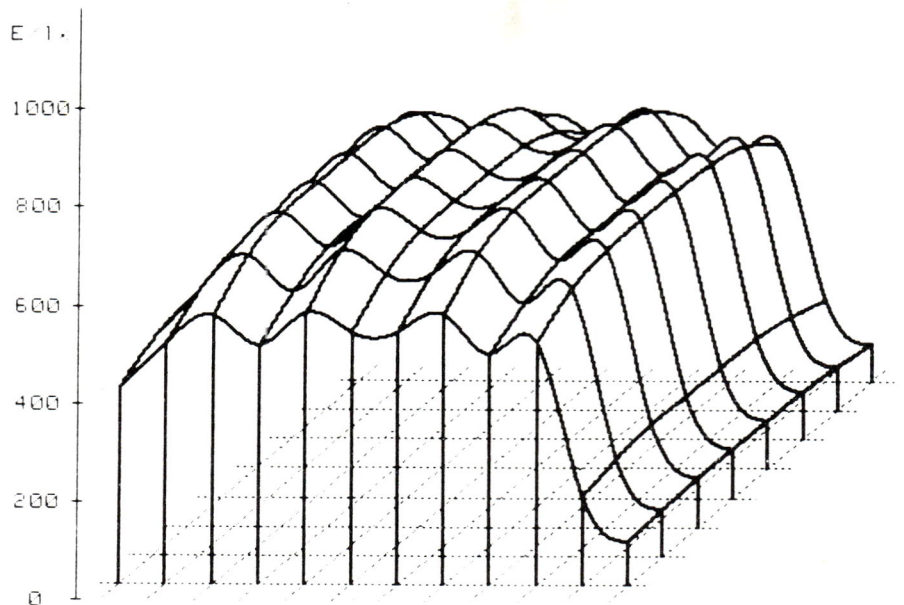

## 4.5.4.4 Area sources

When both the width and length of a luminaire are greater than a fifth of the distance from its centre to the point of illumination, then the source should be considered as an area source.

Area source calculations are by far the most complicated of the three types. There is no simple equivalent of the inverse square law or aspect factor calculation for area sources. Indeed, for many situations the formulae have not been solved. For this reason, and the fact that area source calculations are not often required, only the simple case of a uniform area source is considered in this Code (see appendix 10).

It is important to realise that a luminaire is only designated as a point source or a line source because of its distance from the point of illumination. Thus a 1.8m batten fitting will behave like a line source for points up to 9m away, but, at greater distances, the errors in the inverse square law calculation become acceptable and the luminaire can be treated as a point source.

## 4.5.4.5 Interreflected light at a point

The preceding methods of calculating illuminance at a point deal only with the calculation of direct illuminance and do not allow for interreflected light.

Interreflected light can be dealt with in one of four different ways:

(a) Calculate the interreflected illuminance at the point and add it to the calculated direct illuminance. To do this first calculate the final illuminances of the walls, floor and ceiling (using transfer factors, see CIBS Technical Memorandum 5: The calculation and use of utilisation factors). Then treat each of these room surfaces as if it was an area source of intensity distribution $I(\theta)$:

$$I(\theta) = I \cos \theta, \text{ where } I = \frac{ER}{\pi}$$

where $E$ is the illuminance of the surface, $R$ is its reflectance, and $\theta$ is the angle from the normal to the surface.

The major snag with this is that the calculations involved are quite lengthy, and only really suited to a computer. Another disadvantage is that the result will be somewhat artificial, since local changes in reflectance can make a vast change to the amount of interreflected light.

(b) Ignore the interreflected light and assume that it will be a bonus, increasing the final illuminance. In other words, calculate the worst case. This is a more practical approach, but may not be adequate for some situations.

(c) Ignore the interreflected light at the point, but calculate the average illuminance on the horizontal reference plane (by the lumen method) for the actual reflectances and for a black room (zero reflectances). These two figures, the average final illuminance, which includes interreflection, and the average direct illuminance can be found easily. The difference gives a good indication of the proportion of the total luminous flux that is reflected to the point.

This has the advantage that it is easy to do and is more precise than the second approach. It can also be applied to planes other than the horizontal if the appropriate utilisation factors are available.

(d) Calculate the average indirect illuminance $E_{ind}$ by the formula

$$E_{ind} = \frac{F.DLOR.}{A(F)} \left| \frac{(DR.RE(F)) + ((1 - DR).RE(W)) + (FFR.RE(C))}{2 + \frac{2}{RI}(1 - RE(W)) - RE(C) - RE(F)} \right|$$

where
| | | |
|---|---|---|
| $F$ | = | installed bare lamp luminous flux (lumens) |
| $DLOR$ | = | downward light output ratio |
| $A(F)$ | = | area of floor (metre$^2$) |
| $DR$ | = | direct ratio of installation |
| $FFR$ | = | flux fraction ratio of installation |
| $RI$ | = | room index |
| $RE(F)$, $RE(C)$, $RE(W)$ | = | effective reflectances of floor cavity, ceiling cavity and walls |

This indirect illuminance is assumed to be uniformly distributed over all the room surfaces. It may therefore be added to the direct illuminance calculated at each individual point. This method is recommended when programmable calculators are to be used. It has the merit that it can also be used to show the effect of interreflected light on wall and ceiling illuminance ratios and on the vector/scalar ratio.

## 4.5.5 Glare Index

The CIBS Glare Index system for the evaluation of discomfort glare will be detailed in a forthcoming CIBS publication.

There are two methods of calculation: *(a)* The calculation of glare index for the actual installation using the basic formula (see appendix 5). *(b)* The calculation of glare index from tables based on photometric data for the real luminaire (published by the manufacturer), in the form of an uncorrected glare index with correction factors.

The first method requires the use of a suitable computer and program and the availability of detailed photometric data. The advantages to be gained from the use of a program are that any layout of luminaires seen from any viewing position can be considered. These advantages are seldom sufficient to justify the expense of developing or buying the software. The second method is sufficiently accurate for most purposes and is easy to use. Table 4.1 shows an uncorrected glare index table for a real luminaire. It is calculated from the glare index formula discussed in appendix 5.

The table is based upon a number of assumptions. These are: *(a)* the luminaires are at a spacing to height ratio of 1.0; *(b)* the luminaires are at a height of 2.0m above eye level; *(c)* the total light output of the lamps in the luminaire is 1000 lumens; *(d)* the observer is located at the mid-point of a wall, with a horizontal line of sight towards the centre of the opposite wall; *(e)* the eye level is taken as 1.2m above floor level.

Correction terms can be applied to the uncorrected glare index to allow for changes in mounting height and lamp output per luminaire. At the present time there is no correction for other spacing to height ratios.

Uncorrected glare indices are tabulated according to room dimensions and reflectances. Figure 4.19 shows the method of specifying the room dimensions. The Y dimension is always parallel to the line of sight and the X dimension is perpendicular to the line of sight. They are both expressed as multiples of the mounting height above eye level.

The worst glare condition will occur for viewing from the centre of either the long wall or the short wall. The tables permit either to be calculated by interchanging X and Y.

One view of the room will show the ends of the luminaires (endwise view) and the other view will show the sides (crosswise view). The two halves of the table cater for this.

When the Glare Index has been found (interpolation may be needed) it must be corrected for: *(a)* mounting height above 1.2m eye level if this differs from 2m; *(b)* Total lamp luminous flux per luminaire if this differs from 1000 lm; *(c)* extra correction terms if the published uncorrected glare index table covers a variety of luminaire sizes or lamp types.

These correction terms are added (or subtracted) from the uncorrected glare index to give the final glare index of the installation.

The height correction term and the total luminous flux correction term can be calculated as follows:

Height correction term = $4 \log_{10} H - 1.2$
where $H$ = the height above eye level (m)

Total lamp luminous flux per luminaire correction term = $6 \log_{10} (n.F) - 18$

where $F$ = the luminous flux/lamp (lumens)
$n$ = number of lamps/luminaire

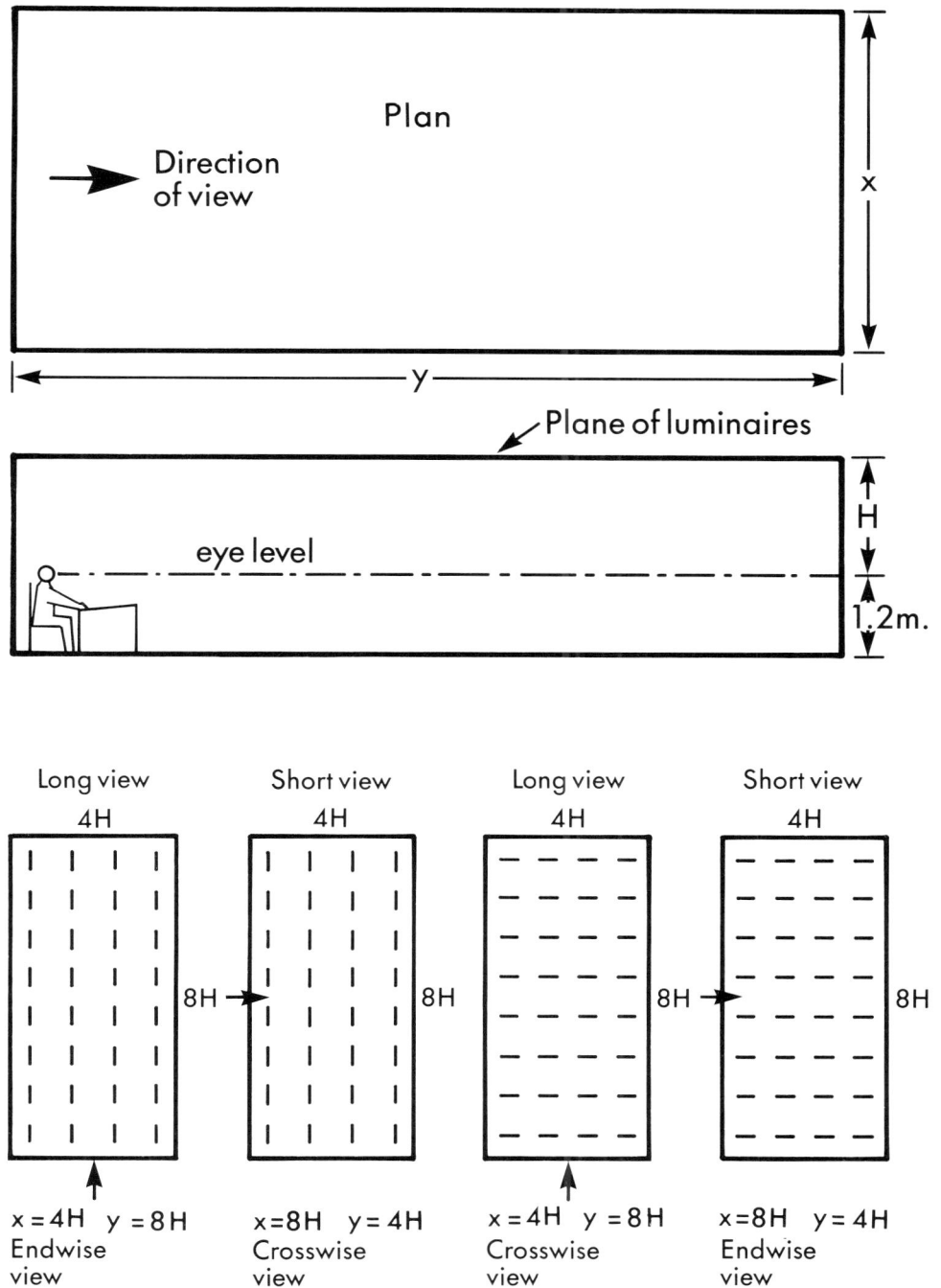

Fig 4.19 Method of specifying room dimensions in accordance with the direction of view for glare index calculations

Long view 4H | Short view 4H | Long view 4H | Short view 4H

8H → | 8H | 8H → | 8H

x = 4H   y = 8H
Endwise view

x = 8H   y = 4H
Crosswise view

x = 4H   y = 8H
Crosswise view

x = 8H   y = 4H
Endwise view

## 4.5.6  Emergency lighting

The Fire Precautions Act, 1971 and the Health and Safety at Work etc. Act of 1974 make it obligatory to provide adequate means of escape in all places of work and public resort. Emergency lighting is an essential part of this requirement. BS 5266, the Code of Practice for the Emergency Lighting of Premises, lays down minimum standards for the design, implementation and certification of emergency lighting installations.

Emergency lighting is lighting provided for use when the main lighting fails for whatever reason. There are two types: escape lighting and standby lighting.

Escape lighting is provided to ensure the safe and effective evacuation of the building. It must: *(a)* indicate clearly and unambiguously the escape routes; *(b)* illuminate the escape routes to allow safe movement towards and out of the exits; *(c)* ensure that fire alarm call points and fire equipment provided along the escape route can be readily located.

# Table 4.1 Typical uncorrected Glare Index Table

| Glare Indices | | | | | | | | | | | |
|---|---|---|---|---|---|---|---|---|---|---|---|
| Ceiling reflectance | | .70 | .70 | .50 | .50 | .30 | .70 | .70 | .50 | .50 | .30 |
| Wall reflectance | | .50 | .30 | .50 | .30 | .30 | .50 | .30 | .50 | .30 | .30 |
| Floor reflectance | | .20 | .20 | .20 | .20 | .20 | .20 | .20 | .20 | .20 | .20 |
| Room dimension X | Y | Viewed crosswise | | | | | Viewed endwise | | | | |
| 2H | 2H | 7.0 | 8.4 | 8.0 | 9.5 | 10.8 | 6.8 | 8.2 | 7.8 | 9.2 | 10.5 |
| | 3H | 8.9 | 10.2 | 10.0 | 11.3 | 12.6 | 8.6 | 9.8 | 9.6 | 10.9 | 12.2 |
| | 4H | 9.9 | 11.1 | 10.9 | 12.2 | 13.5 | 9.4 | 10.6 | 10.4 | 11.7 | 13.0 |
| | 6H | 11.0 | 12.1 | 12.0 | 13.2 | 14.5 | 10.3 | 11.4 | 11.3 | 12.5 | 13.8 |
| | 8H | 11.6 | 12.6 | 12.6 | 13.7 | 16.1 | 10.7 | 11.8 | 11.7 | 12.9 | 14.2 |
| | 12H | 12.2 | 13.2 | 13.2 | 14.3 | 16.7 | 11.1 | 12.1 | 12.1 | 13.2 | 14.6 |
| 4H | 2H | 7.7 | 8.9 | 8.7 | 10.0 | 11.3 | 7.5 | 8.7 | 8.5 | 9.3 | 11.1 |
| | 3H | 10.0 | 11.0 | 11.0 | 12.1 | 13.5 | 9.6 | 10.6 | 10.9 | 11.7 | 13.1 |
| | 4H | 11.2 | 12.1 | 12.2 | 13.2 | 14.6 | 10.6 | 11.6 | 11.7 | 12.7 | 14.1 |
| | 6H | 12.5 | 13.4 | 13.6 | 14.5 | 15.9 | 11.8 | 12.6 | 12.3 | 13.7 | 15.1 |
| | 8H | 13.3 | 14.0 | 14.4 | 15.2 | 16.6 | 12.3 | 13.1 | 13.4 | 14.2 | 15.7 |
| | 12H | 14.0 | 14.3 | 15.1 | 15.9 | 17.3 | 12.8 | 13.5 | 13.9 | 14.7 | 16.1 |
| 8H | 4H | 11.8 | 12.5 | 12.9 | 13.7 | 15.2 | 11.4 | 12.2 | 12.5 | 13.3 | 14.7 |
| | 6H | 13.5 | 14.2 | 14.6 | 15.3 | 16.8 | 12.8 | 13.5 | 13.9 | 14.6 | 16.1 |
| | 8H | 14.4 | 15.0 | 15.5 | 16.1 | 17.6 | 13.5 | 14.1 | 14.6 | 15.2 | 16.7 |
| | 12H | 15.4 | 16.0 | 16.6 | 17.1 | 18.6 | 14.2 | 14.8 | 15.4 | 15.9 | 17.4 |
| 12H | 4H | 12.0 | 12.7 | 13.1 | 13.8 | 15.3 | 11.6 | 12.3 | 12.7 | 13.5 | 14.9 |
| | 6H | 13.7 | 14.3 | 14.9 | 15.5 | 17.0 | 13.1 | 13.7 | 14.3 | 14.9 | 16.4 |
| | 8H | 14.8 | 15.3 | 16.0 | 16.5 | 18.0 | 14.0 | 14.5 | 15.1 | 15.7 | 17.2 |
| | 12H | 15.7 | 16.2 | 16.8 | 17.3 | 18.8 | 14.6 | 15.0 | 15.7 | 16.2 | 17.7 |

Some building areas cannot be evacuated immediately in the event of an emergency or power failure. This is often because life would be put at risk; for example, in a hospital operating theatre, or in some chemical plants where shut-down procedures must be used. In these circumstances, appropriate activities must be allowed to take place and standby lighting is required. The level of standby lighting will depend upon the nature of the activities, their duration and the associated risk.

Standby schemes may have to provide from 5% to 100% of the design service illuminance according to circumstances. In some cases the lighting requirements may be more demanding in the event of power failure. Correct lighting requirements can only be established by careful analysis.

Standby lighting can be regarded as a special form of conventional lighting and dealt with accordingly. Escape lighting requires different treatment.

## 4.5.6.1 Escape lighting requirements

BS 5266, like most British Standards, is not a legal requirement. However, it can acquire legal status by being adopted as part of the local bye-laws. Although most enforcing authorities quote BS 5266, many modify the conditions. For example, they may insist on a higher illuminance. In addition to this legal requirement, many organisations have their own more onerous standards. Therefore, the designer must thoroughly investigate the standards that apply to the building.

## 4.5.6.2 Marking the route

All exits and emergency exits must have exit or emergency exit signs. Where direct sight of an exit is not possible, or there could be doubt as to the direction, then direction signs with an appropriate arrow and the words exit or emergency exit are required. The idea is to direct someone who is unfamiliar with the building to the exit. All of these signs must be illuminated at all reasonable times so that they are legible.

*Detailed planning*                    *CIBS Code 1984*

## 4.5.6.3 Illuminating the route

The minimum illuminance along the centre line of a clearly defined escape route should be 0.2 lx and 50% of the route width for escape routes up to 2 metres wide should be lit to a minimum of 0.1 lx. These criteria are sufficient for most interiors but for auditoria, where many people may be assembled, more elaborate criteria are required. For an auditorium having fixed seating, the average horizontal illuminance on a plane 1 m above the floor/pitch line should not be less than 0.1 lx over any seated area. Gangways should be lit as for clearly defined escape routes. For auditoria where adaptable seating is used the average horizontal illuminance on an unobstructed floor should not be less than 0.2 lx. For auditoria which are permanently constructed as a series of steppings, the average horizontal illuminance in front of each unobstructed step should not be less than 0.2 lx. Where the escape route is not defined, the average horizontal illuminance should not be less than 0.2 lx. It is important to note that most fires are accompanied by smoke and that smoke may affect the illuminances produced on the escape routes.

The emergency lighting must reach its required illuminance 5 seconds after failure of the main lighting system. If the occupants are familiar with the building, this time can be increased to 15 seconds at the discretion of the enforcing authority.

Other important factors are:

*Glare* – The emergency lighting luminaires should not cause problems of disability glare. Luminaires should be mounted at least 2m above floor level in order to avoid glare but should not be too high or they may become obscured by smoke.

*Exits and changes of direction* – Luminaires should be located near each exit door and emergency exit door and at points where it is necessary to emphasise the position of potential hazards, such as changes of direction, staircases, changes of floor level and so on.

*Fire equipment* – Fire fighting equipment and fire alarm call points along the escape route must be adequately illuminated at all reasonable times.

*Lifts and escalators* – Although these may not be used in the event of fire they should be illuminated. Emergency lighting is required in each lift car in which people can travel. Escalators must be illuminated to the same standard as the escape route to prevent accidents.

*Special areas* – Emergency lighting luminaires are required in all control rooms and plant rooms. In toilets of over 8m$^2$ gross area, emergency lighting should be installed to provide a minimum of 0.2 lux.

## 4.5.6.4 Systems

The design aspects of emergency lighting will be discussed in detail in a forthcoming CIBS publication. In summary, there are two main supply systems, generators and batteries. Generators must provide the required illuminance within 5-15 seconds, the actual times being at the discretion of the enforcing authority. Therefore, they must either be running continuously or automatically start within the maximum time allowed. If this cannot be achieved then auxiliary battery systems must be used.

Generators require considerable capital investment and are difficult to justify except for standby systems on large sites.

Battery systems can be of two types: central systems, where the batteries are in banks at one or more locations; and self-contained systems, where each individual luminaire has its own battery. Central systems have battery rooms or cubicles in which the charger, batteries and switching devices are located.

Battery cubicles can be designed to simplify system maintenance. However, the system must be well designed if it is not to be rendered inoperative if damaged by fire.

Self-contained luminaires are self-powered and operate independently in an emergency. Thus, although an individual luminaire may be destroyed in a fire, the other luminaires will be unaffected. The fact that each luminaire is an independent unit means that maintenance must be thorough. For most applications self-contained luminaires must operate for a period of 1 hr to 3 hrs. Many designers base their designs on the 3 hr standard because it gives greater reliability. Self-contained luminaires can have three modes of operation:

*Maintained* – In this the lamp is on all the time. Under normal conditions it is powered directly or indirectly by the mains. Under emergency conditions it uses it's own battery supply.

*Non-maintained* – In this the lamp is off when mains power is available to charge the batteries. Upon supply failure the lamp is energised from the battery pack.

*Sustained* – This is a hybrid of the previous two. A lamp is provided which operates from the mains supply under normal conditions. Under emergency conditions a second lamp, powered from the battery pack, takes over. Sustained luminaires are often used for exit signs.

Systems of self-contained luminaires are the easiest and most flexible to install but their effective life is likely to be less than that of central battery systems. Also, maintenance and testing must be thorough if operation in the event of emergency is to be guaranteed.

### 4.5.6.5 Calculating the illuminance along the escape route

In order to ensure that the minimum illuminances are within the prescribed limits, the calculation of illuminance along the escape route is required. Methods and formulae for doing this are given in section 4.5.4. It is important to base the calculations upon realistic photometric data for the luminaire and lamps. It is essential that the calculations portray the worst set of conditions that are likely to be encountered; the luminaires may be at the end of their cleaning and maintenance cycle; the lamps may be at the end of their useful life; the batteries may be at the end of their discharge period; the ambient temperature may be excessive; and so on. These are just examples, and the true worst condition must be determined.

Many manufacturers provide design information to help in planning emergency lighting. It is important to use this data correctly and with the appropriate value of 'emergency lighting design lumens' for the lamp.

### 4.5.6.6 Planning sequence

When planning an emergency lighting system the following sequence will help.

*(1)* Define the exits and emergency exits.
*(2)* Mark the escape routes.
*(3)* Identify any problem areas. For example, areas that will contain people unfamiliar with the building, plant rooms, escalators, etc.
*(4)* Mark the location of exit signs. These can be self-illuminated or illuminated by emergency lighting units nearby. Mark these onto the plan.
*(5)* Where direction signs are required mark these and provide necessary lighting.
*(6)* Identify the areas of the escape route illuminated by the lighting needed for the signs.
*(7)* Add extra luminaires to complete the lighting of the escape route, paying particular attention to stairs and other hazards. Remember to

*Detailed planning*

allow for shadows caused by obstructions or bends in the route.

(8)  Add extra luminaires to satisfy the problem areas identified in item 3 of this sequence. Make sure that lighting outside the building is also adequate for safe evacuation.

(9)  Check that all fire alarm call points and fire equipment have been adequately dealt with.

## 4.5.7 Checklist

At some point, the designer should check systematically that he has taken account of all the factors relevant to the design of the lighting installation. The following checklist is given as an aid to this process. The checklist is in the form of a series of headings and subsidiary questions. The headings indicate the area to be considered and the questions refer to the most commonly occurring aspects of the topic. In any specific situation there may be questions other than those listed which need to be considered under the general headings.

### Objectives

*Safety requirements* – What hazards need to be seen clearly, what form of emergency lighting is needed, is a stroboscopic effect likely?

*Task requirements* – What are the tasks to be performed in the interior, where are the tasks, what planes do they occupy? what aspects of lighting are important to the performance of these tasks, are optical aids necessary?

*Appearance* – What impression is the lighting required to create?

### Constraints

*Statutory* – Are there any statutory requirements which are relevant to the lighting installation.

*Financial* – What is the budget available, what is the relative importance of capital and running costs?

*Physical* – Is a hostile and/or hazardous environment present so special equipment is required, are high or low ambient temperatures likely to occur, is noise from control gear likely to be a problem, are mounting positions restricted, is there a limit on luminaire size?

*Historical* – Is the choice of equipment restricted by the need to make the installation compatible with existing installations?

### Specification

*Source of recommendations* – What is the source of the lighting recommendations used, how authoritative is this source?

*Form of recommendations* – Have all the relevant lighting variables been considered, e.g. design service illuminance, uniformity, illuminance ratios, surface reflectances and colours, light source colour, colour rendering group, limiting glare index, vector/scalar ratio, veiling reflections?

*Qualitative requirements* – Have the aspects of the design which cannot be quantified been carefully considered?

### General Planning

*Natural and electric lighting* – What is the relationship between these forms of lighting, is it possible or desirable to provide a control system to match the electric lighting to the daylight available?

*Protection from solar glare and heat gain* – Are the windows designed to limit solar glare and heat gain to the occupants of the building, do the window walls have suitable reflectances?

*Choice of electric lighting system* – Is general, localised or local lighting most appropriate for the situation, does obstruction make some form of local lighting inevitable?

*Choice of lamp and luminaire* – Does the light source have the required lumen output, luminous efficacy, colour properties, life, run up and restrike properties; is the proposed lamp/luminaire package suitable for the application, will the luminaire be safe in the environmental conditions, will it withstand the environmental conditions, does it conform to BS 4533 or other appropriate standard, does the luminaire have an appropriate appearance and will it enable the desired effect to be created. Is reliable photometric data available?

*Maintenance* – Has a realistic light loss factor been estimated, has a maintenance schedule been agreed, is the equipment resistant to dirt deposition, can the equipment be easily maintained, is the equipment easily accessible, will replacement parts be readily available?

*Control systems* – Are control systems for matching the operation of the lighting to the availability of daylight and the level of occupancy appropriate? Is a dimming facility desirable, are any manual switches easily accessible and is their relationship to the lighting installation understandable?

*Interactions* – How will the lighting installation influence other building services? Is it worth recovering the heat produced by the lamps?

## Detailed planning

*Layout* – Is the layout of the installation consistent with the objectives and the physical constraints? Has allowance been made for the effects of obstruction by building structure, other services, machinery and furniture? Has the possibility of undesirable high luminance reflections from specular surfaces been considered; does the layout conform to the spacing height ratio criteria?

*Mounting and electrical supply* – How are the luminaires to be fixed to the building, what system of electricity supply is to be used, does the electrical installation comply with the latest edition (with any amendments) of the IEE Regulations for Electrical Installations?

*Calculations* – Have the design service illuminances been calculated for appropriate planes, have the most suitable calculation methods, e.g. lumen method, illuminances for point, line or area sources, been used? Has the glare index been calculated? Has up to date and accurate photometric data been used.

*Verification* – Does the proposed installation meet the specification of lighting conditions? Is it within the financial budget? Is the installed efficacy within the recommended range? Does the installation fulfill the design objectives?

## 4.5.8 Statement of assumptions

When submitting a design proposal to clients, it will usually be necessary to supply information on the following topics: *(a)* the design specification, i.e. the type of lighting system, the design service illuminance, the glare index, the lamp colour properties the wall/task illuminance ratio, the ceiling/task illuminance ratio, the vector/scalar ratio, the installed efficacy, the operating efficacy and other criteria as applicable; *(b)* the equipment to be used, e.g. lamps, luminaires, control systems; *(c)* the layout of the equipment; *(d)* the costs, in an appropriate form; *(e)* the lighting conditions which will be achieved; *(f)* the installed efficacy and operating efficacy of the installation; *(g)* the assumptions made in the design.

The level at which each of these topics is covered is a matter of commercial judgement. Unfortunately such judgement sometimes leads to ambiguity in the information supplied to the client, particularly regarding the lighting conditions which will be achieved and the assumptions made in the design. If the client is to compare design proposals on an equitable basis it is essential that the assumptions made in the design are given for each aspect of the lighting conditions. Table 4.2 lists the assumptions that are usually involved in the estimation of the lighting conditions achieved by a general lighting installation. If localised lighting is being proposed, it will also be necessary to state the areas to which each illuminance applies and to give details for each area separately. If local lighting is being proposed, it will be necessary to give details of the general surround illuminance and the task illuminance, the latter being divided into the contributions from the local luminaire and from the general surround lighting. Special situations may involve additional assumptions, in which case these too should be stated in the design proposal.

**Table 4.2   Assumptions\* which should be made explicit when describing the lighting conditions that will be produced by a proposed general lighting installation**

| Lighting Condition | Assumptions that need to be stated |
| --- | --- |
| Initial illuminance | Room Index, effective reflectance of ceiling cavity, effective reflectance of walls and effective reflectance of floor cavity used in establishing the Utilisation Factor; the initial luminous flux of the lamp used |
| Illuminance at a specified time | As for initial illuminance, plus the elapsed time for which the illuminance is given, and Light Loss Factor (see below) |
| Glare Index | Calculation method, viewing position |
| Wall/task illuminance ratio<br>Ceiling/task illuminance ratio<br><br>Vector/Scalar ratio | If using the illuminance ratio charts then<br>  – ceiling wall and floor reflectance classes<br>  – Room Index, BZ class of luminaire<br>    flux fraction ratio of luminaire |
| Light Loss Factor | Elapsed time for which Light Loss Factor is given, Lamp Lumen Maintenance Factor and hours of operation of lamps, the Luminaire Maintenance Factor and luminaire cleaning schedule, Room Surface Maintenance Factor and room cleaning and painting schedule |
| Installed efficacy | Effective reflectance of ceiling cavity, effective wall reflectance, effective reflectance of floor cavity and Room Index used in establishing the Utilisation Factor; the luminous efficacy of the lamp used (including circuit watts). |
| Operating efficacy | Maximum hours of use and hours of equivalent full installation use assumed in the calculation of load factor |

*\*Assumptions may be made by the designer or by the client and given to the designer in the form of a specification.*

# 4.6 Appraisal

After an installation has been completed it is often instructive for the designer to undertake an appraisal. In addition to the obvious subjective assessment made by the designer, a completed appraisal should involve: a photometric survey of the lighting conditions achieved by the installation; a discussion with the clients centred on their assessment of the installation; a discussion with the users of the installation to discover their reactions to the installation.

The results of the photometric survey can be compared with the quantitative elements of the specification. Hence the extent to which the installation meets the specification can be established (appendix 11 describes the important aspects of field measurements of lighting). The discussion with the clients and the users of the installation should reveal the extent to which the installation meets their expectations and requirements, although it may tell the designer more about the limitations of the original design objectives and specification than of the design itself. The justification for undertaking an appraisal is reassurance and education.

# APPENDIX 1 ALTERNATIVE MEASURES OF ILLUMINANCE

## A1.1 Introduction

The standard service illuminances recommended in this Code are to be provided on a plane appropriate for the application. This plane may be horizontal, vertical or inclined. However, there will be some applications for which either: (a) there are a large number of different planes which all have to be lit adequately by the same lighting installation; or (b) the planes of interest are not defined for the lighting designer or are likely to change rapidly with time; or (c) there are no obviously important planes.

In any of these circumstances the usual procedure is to assume the important plane is horizontal and design to provide the design service illuminance on this horizontal plane. For most cases this will give a satisfactory result because the inter-reflection of light in the space ensures that the illuminance of vertical and inclined planes is a reasonable proportion of the illuminance on the horizontal plane, typically greater than 0.4. However, where the inter-reflected proportion of the luminous flux in the interior is low, then this assumption is no longer valid. The inter-reflected component can be low either because there is considerable obstruction by equipment in the space or because the room surface reflectances are low, or because the luminous intensity distribution of the luminaire is strongly directional, usually vertically downward. In these circumstances the designer can use one of three approaches to check that the installation proposed will give a satisfactory illuminance on planes other than the horizontal. These approaches are: (a) calculate the illuminances on appropriate vertical planes; (b) calculate the mean scalar illuminance; (c) calculate the mean cylindrical illuminance.

## A1.2 Vertical Illuminance

For all installations, the illuminance on a specific vertical surface can be calculated by a point by point method (see appendix 10), but for a regular array of symmetrical luminaires a simpler method is available. The average wall/task illuminance ratio due to a regular array of symmetrical luminances is virtually independent of the room index and depends principally on the luminaire and on surface reflectances. This means that the average illuminance on vertical surfaces should be approximately the same as the average wall illuminance. This is found from the expression

Average wall illuminance = average horizontal illuminance × wall/task illuminance ratio

The average horizontal illuminance is normally found by the lumen method; the wall/task illuminance ratio is found from the appropriate IR Chart in CIBS (IES) Technical Report No. 15[1].

## A1.3 Scalar illuminance

The scalar illuminance is the average illuminance over the surface of a very small sphere at a point[2]. It is approximately related to the illuminance on a horizontal plane by the formula

$$\overline{E}_s = E_h. \; (1 - RE(F)). \; E_s/\vec{E}$$

where
$\overline{E}_s$ = mean scalar illuminance (i.e. the mean value of the scalar illuminance averaged over a horizontal reference plane)

$E_h$ = illuminance on the horizontal reference plane

$RE(F)$ = floor cavity reflectance

$\vec{E}/E_s$ = vector/scalar ratio estimated from the appropriate IR Chart (see CIBS (IES) Technical Report No. 15.[1])

As scalar illuminance is entirely independent of the direction from which the light is incident, it is a very appropriate measure for use in interiors where there are no obviously important planes. A hotel lobby is an example of such an interior. For a satisfactory appearance, the mean scalar illuminance should be at least 40% of the illuminance on the horizontal plane. Methods of measuring scalar illuminance in the field are discussed in Appendix 11.

## A1.4 Mean cylindrical illuminance

The mean cylindrical illuminance is the average illuminance over the surface of a very small cylinder at a point, the axis of the cylinder being vertical[3]. It is approximately related to the mean scalar illuminance and the illuminance on a horizontal plane by the equation

$$\overline{E}_c = 1.5 \, \overline{E}_s - 0.25 \, E_h \, (1 + RE(F))$$

where
$\overline{E}_c$ = average mean cylindrical illuminance (i.e. the mean cylindrical illuminance averaged over a horizontal reference plane)

other terms are as defined above

The mean cylindrical illuminance is most sensitive to the illuminance on vertical planes and is insensitive to illuminance on horizontal planes. Therefore it can most appropriately be used where vertical or near vertical planes are likely to be important. This is the situation in much of industry[4]. For a satisfactory installation mean cylindrical illuminance should be at least 0.4 of the horizontal illuminance. Methods whereby mean cylindrical illuminance can be measured in the field are discussed in appendix 11.

## A1.5 References

1 Illuminating Engineering Society, Technical Report 15, Multiple criterion design, a design method for interior electric lighting installations, CIBS, London, 1977.
2 Lynes, J.A. Burt, W. Jackson, G.K. and Cuttle, C.C. The flow of light in buildings, Trans. Illum. Engng. Soc. (London), 31, 65, 1966.
3 Epaneshnikov, M.M., Obrosova, N.A., Sidorova, T.N. & Undasynov, G.N. New characteristics of lighting conditions for the appearance of public buildings and methods for their calculation. Proc. CIE 17th Session, Barcelona, 1971.
4 Carlton, J.W. Effective use of lighting, in Developments in Lighting 2, edited by D.C. Pritchard, Applied Science Publishers, London, 1982.

# APPENDIX 2   VISUAL INSPECTION

## A2.1   Introduction

Visual inspection of products is an important aspect of many manufacturing processes. The accuracy of visual inspection is influenced by four aspects of the situation: the people, the task, the environment and the organisation[1]. The people doing the inspecting are important because their visual capabilities affect the ease with which they can see the features of interest and their experience of the work will influence the way that they examine the product. The task is important in the way it is presented to the inspector. It can be presented for different periods of time, regularly or irregularly, stationary or moving, in ordered groups or individually, with different types of features occurring with different probabilities. The environment, which includes the lighting, is important because it affects the conspicuity of the features of interest. The organisation is important because it includes such aspects as the number and duration of rest pauses, the understanding the inspector has of what constitutes a feature that should be detected, the extent to which the inspector is kept informed of his performance and the relationship between the inspection and production sides of the business.

Thus the lighting of the product is only one of several factors influencing the accuracy of visual inspection; but it is an important factor and often it is one of the easiest to change. All too frequently the only concession made to the visual difficulty of inspection work is to provide a high illuminance in the inspection area. Although this is generally helpful, special lighting designed to reveal the critical features can be much more effective[2,3,4,5].

## A2.2   Principles of Inspection Lighting

The aim of all inspection lighting is to increase the conspicuity of the features of the product that determine whether it is acceptable or not. There are a number of lighting techniques that can be used to achieve this aim, the choice between them depending on the nature of the material from which the product is made and the type of feature being sought. An outline of the main techniques is given below.

### A2.2.1 Controlling the direction of incident light

For flat materials with diffuse reflection characteristics where the features being sought damage the surface, e.g. scratches on a tile, the most effective technique is to light the material at a glancing angle so that no light is reflected towards the inspector's eyes. The damage to the surface will be emphasised by the highlights and shadows created around it. Figure A2.1 shows the effect of glancing angle lighting on the visibility of damage in cloth. It should be noted that glancing angle lighting is very effective for revealing any form of texture on a surface, although whether this is an advantage or not depends on the particular circumstances; too much detail can be confusing.

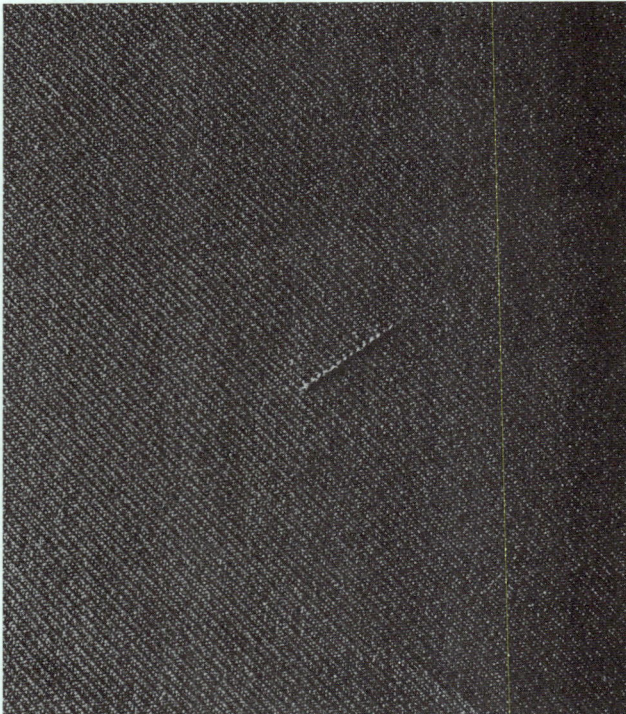

**Fig. A2.1**   Some damaged material lit by very directional lighting

**Fig. A2.2**   A scribed cross on an aluminium sheet lit by a luminaire reflected away from the observer

*Appendix 2 Visual inspection*

For materials with specular reflection properties where the features being sought cause a change in the specular reflection pattern of the surface e.g. scribe marks on stainless steel, there are two useful lighting techniques involving reflection. The first requires the reflection of a luminaire away from the observer. Then the damaged area will appear as bright on a dark background. The second requires the reflection of a luminaire towards the observer. Then the damaged area will appear as dark on a bright background. These differences occur in both cases because the damaged area scatters the light rather than reflecting it specularly. Figure A2.2 shows this technique being used for the inspection of scratches on polished metal.

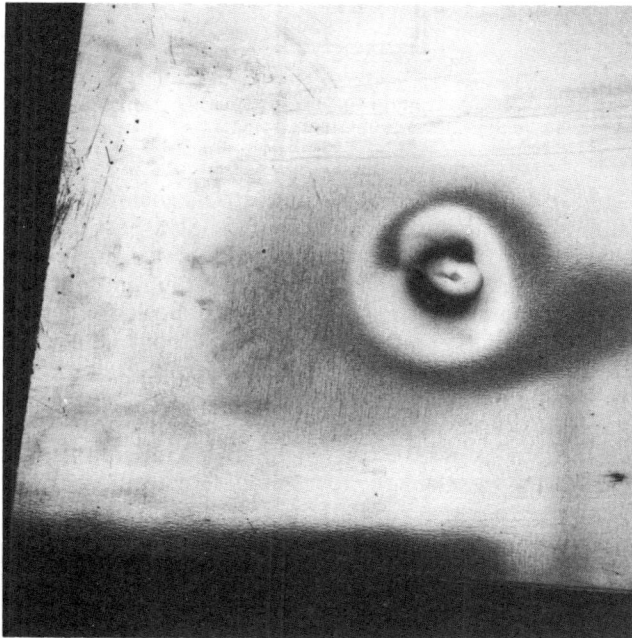

**Fig. A2.3** Departures in form revealed by non-uniformities in the luminance of the reflected image of a luminaire

### A2.2.2 Reflection of a large area of low luminance

For materials with specular reflection properties but where the feature being sought does not change the reflection characteristics but only the form of the product, e.g. a dent in a silver plated surface, a useful technique is to reflect a large area of low luminance from the product towards the inspector. Then, departures in form are revealed by non-uniformities in the luminance of the reflected image. The luminous area which is reflected from the product should be large enough to cover the whole product when viewed from the inspector's position.

This technique is quite effective for dents which have a small radius of curvature (Fig A2.3). However, where the dents have a large radius of curvature it will be advantageous to superimpose a regular grid over the large area of luminance. Then dents in the products will be shown by distortions in the reflected image of the grid.

With or without a grid, this technique depends on a non-uniformity in the image reflected from the product. Therefore this technique is only really suited for flat objects, more complex shapes producing too complicated a pattern of non-uniformities.

It should be noted that this technique of reflecting a large area of low luminance from a specular surface can be useful when it is necessary to separate a specular surface from a diffusely reflecting surface. For example, for examining the continuity of printed circuits, the reflection of a large area of low luminance towards the observer increases the conspicuity of the printed circuit from its background and reveals any breaks in it (Fig A2.4).

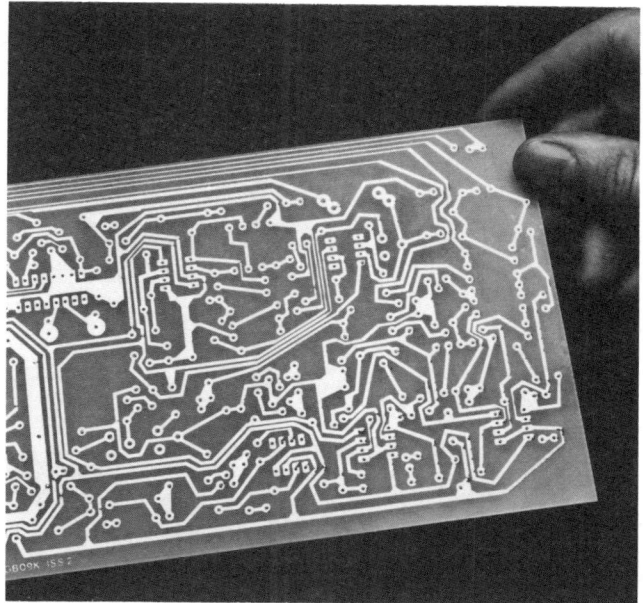

**Fig. A2.4** Reflection of a large area of low luminance emphasises the circuit on a circuit board

### A2.2.3 Transmission of light

*Transmission of a large area of low luminance.* For transparent and translucent materials where the features being sought are within the material, an effective technique is for the inspector to view a large area of low luminance through the object. Any bubbles, cracks, chips or foreign bodies will create non-uniformity in the transmitted luminance. Again the area of the low luminance should be sufficient to cover the product from the inspector's position.

If the features being sought involve a change in form of the object it is useful to have a regular grid superimposed over the large area of low luminance. Then any changes in form create distortions in the regular grid as seen by the inspector (Fig A2.5). Again this method is suitable only for objects of simple shape; complex shapes are likely to produce many distortions in the grid which will be confusing.

*Oblique transmission.* For transparent materials, where the features being sought are bubbles, cracks, scratches, etc., another useful technique is to light the object with a strongly directional beam into the edge of the product so that no light reaches the inspector directly. Faults in the material will tend to produce scattered light which, when the object is viewed against a dark background, will appear as sparkling highlights. Figure A2.6 shows this technique being used for the inspection of moulded translucent glass.

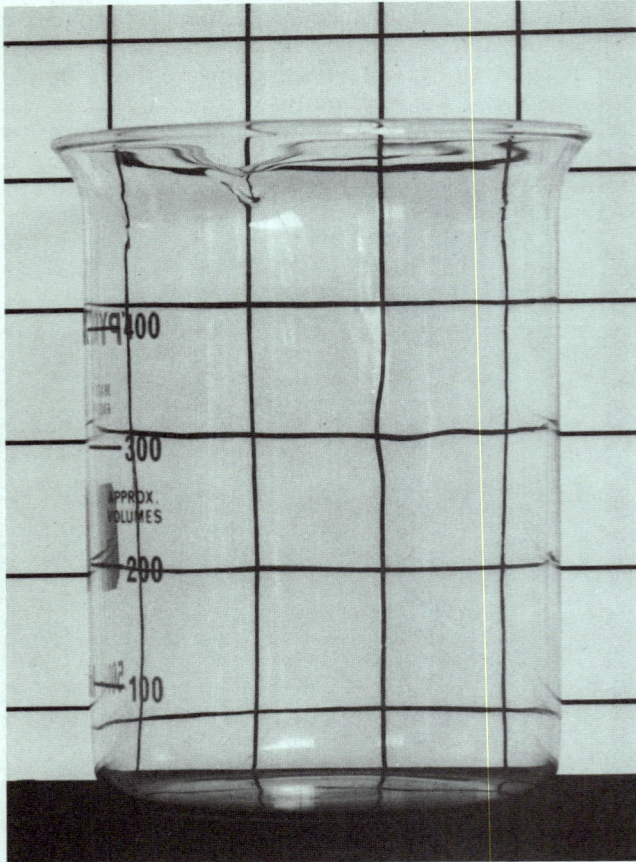

**Fig. A2.5** Defect revealed by distortion of a background grid

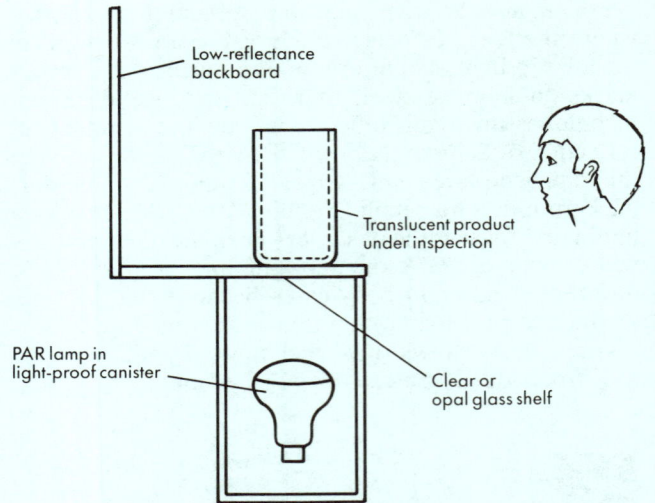

**Fig. A2.6** Oblique lighting aids the inspection of translucent products, such as glass containers

### A2.2.4 Spectral composition of light

*Colour appraisal.* For many products, a common form of visual inspection is the examination of the colour of the product. For example, one of the criteria used to grade agricultural and horticultural produce is their colour. This form of visual inspection is called colour appraisal. The most important lighting variable for colour appraisal work is the spectral composition of the light source used. The lighting designer has two conditions to consider, *(a)* the light source under which the product will be used, *(b)* the light source under which the product will be sold. When these two light sources have similar effects on the colour of the product then the product need only be examined under one light source. However, if the using and selling conditions are markedly different, then inspection under both light sources may be necessary. There is no single light source which is best for colour appraisal work; the choice of light source is determined by the characteristics of the product and the colour appearance that is desired. For this reason, there is no substitute for practical experience when selecting appropriate light sources, although, in general, light sources with high CIE General Colour Rendering Indices enable inspectors to make finer discriminations between hues[6,7]. For colour appraisal work the inspection area should be lit uniformly, without veiling reflections and to the recommended illuminance. The surroundings should be neutral in colour and of medium reflectance.

*Colour matching.* For some industries, such as printing and textiles, an important form of visual inspection is the matching of the colour of a product to an existing standard. Again the most important lighting variable available to the lighting designer is the spectral composition of the light source. BS 950, Artificial daylight for the assessment of colour[8] gives recommendations on the spectral composition of light sources suitable for accurate colour matching in various industries. In general, light sources with high CIE General Colour Rendering Indices[7] are most suitable for accurate colour matching.

For colour matching work the inspection area should be lit uniformly, without veiling reflections and to the recommended illuminance. The surroundings should be neutral in colour and of medium reflectance[8]. Figure A2.7 shows a typical inspection booth used for colour matching work.

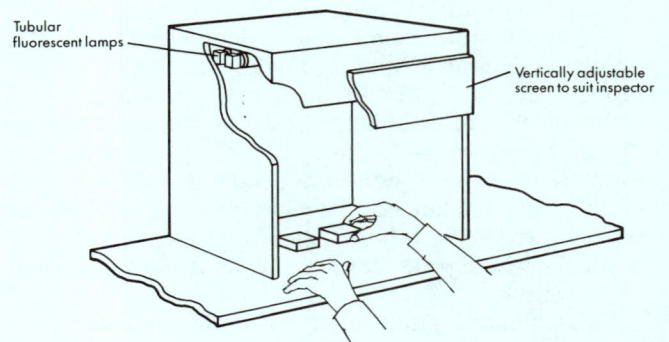

**Fig. A2.7** An inspection booth for colour matching

A special aspect of colour matching is metamerism. This is the phenomenon whereby the colours of two products match under one light source but do not match under another. When this is likely to be important a check for metamerism can be carried out by examining the match between products under two light sources with widely different spectral compositions.

### A2.2.5 Special techniques

So far the lighting techniques considered have been those which use conventional lighting equipment to light the product and the naked eye to examine it. However, there are some inspection techniques which require unusual lighting equipment and/or some form of optical aid.

*Polarization.* Polarized light can be used to examine the stresses within transparent products such as blown glass and moulded plastic. Light from an incandescent lamp is first polarized then transmitted through the product and finally analysed by another polarizer. Stress in the transparent product changes the transmission pattern. Special apparatus is required for this type of work and the equipment usually includes the necessary lighting.

*Fluorescence.* Many materials, including some lubricating and cutting oils, flouresce in the presence of ultra-violet radiation, i.e. they produce visible light. This can be a useful aid in inspection. For example, by using a flourescing dye, the security of a seal can be checked with ultra-violet radiation. Another possibility is to coat the surface of the product with a fluorescing material. Then when illuminated by ultra-violet radiation any hole will appear as black. To use this technique a suitable combination of fluorescing agents and an ultra-violet radiation source is required. Whenever ultra-violet radiation is used it is essential to consider precautions against excessive exposure of eyes or skin[9] (see appendix 8).

*Magnification.* Some productsd are too small to be inspected by the naked eye. For such products some form of magnification is needed. This magnification can be achieved directly by viewing through a microscope or a magnifier. In either case the lighting required can be an intergral part of the equipment provided.

*Stroboscopy.* Sometimes it is necessary to inspect moving parts whilst they are moving rapidly. Stroboscopic illumination can be used to apparently 'slow down' or 'stop' the movement of constant speed rotating or reciprocating machinery. To be successful the technique requires a very regular movement and very little illumination apart from that provided by the stroboscope. For safety reasons, whenever using this technique the operator's attention should be drawn to the fact that although the machine is apparently moving only slowly or has stopped, it is actually still moving at the original speed.

### A2.3 Practical aspects of inspection lighting

#### A2.3.1 Multiple defects

The vast majority of visual inspection work involves looking for more than one feature in a product at a time. As different features may be most effectively revealed by different lighting techniques, this implies that visual inspection should be arranged sequentially with different lighting being produced for each stage. Rarely is this an economic proposition. Usually a compromise is required in which several different forms of lighting are provided at the inspection area and the inspector uses them as he thinks fit[10]. This compromise approach will be most easily accomplished if the product is sufficiently small and light for the inspector to manipulate it amongst the various lighting conditions himself. If this is not possible then sequential inspection lighting based on the most important features which need to be detected may have to be used.

#### A2.3.2 Separation from general lighting

Most inspection lighting techniques rely for their effect on creating conditions very different from those produced by the uniform lighting commonly provided in production areas. Therefore, if inspection lighting is to be effective, it either has to be much brighter than the production area lighting or it has to be separated from the production area lighting. A separate inspection room is rarely possible, but an inspection booth of some sort usually is. Figure A2.8 shows the design of an inspection booth for inspecting veneer pressings for flatness by using glancing angle lighting. Note the canopy shielding the inspector from the overhead lighting.

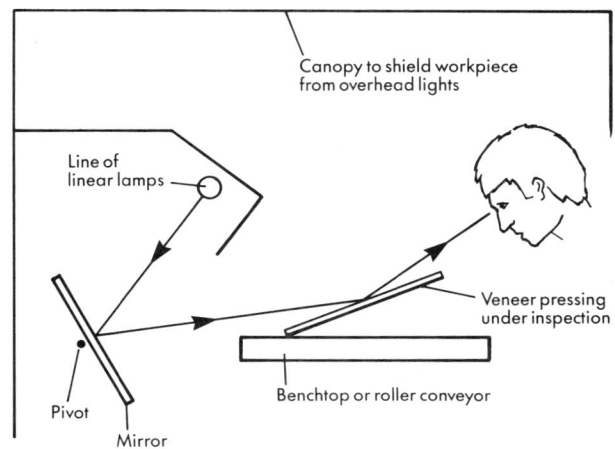

**Fig. A2.8** Directional lighting apparatus for examining veneer pressings for flatness. The linear lamps may be fluorescent tubes with reflectors or double-ended clear tubular filament lamps. The mirror is pivoted to enable it to be set at a convenient angle for the inspector

#### A2.3.3 The need for direct involvement

The most suitable form of inspection lighting for any specific product cannot be determined remotely. The variety of lighting effects that are possible is so large that a simple verbal description of the object to be inspected and the features to be found will be insufficient to identify the appropriate lighting technique. There is no substitute for the lighting designer being directly involved with the inspection task or at the very least, having available samples of the objects to be inspected which contain a representative collection of the features that need to be identified. Only then will the lighting designer be able to 'tailor' the lighting to the task.

### A2.4 REFERENCES

1    Megaw, E.D. Factors affecting visual inspection accuracy. Applied Ergonomics, **10**, 27, 1979.
2    Faulkner, T.W. & Murphy, T.J. Lighting for difficult visual tasks. Human Factors,**15**, 149, 1973.
3    Bellchambers, H.E. & Philipson, S.M. Lighting for inspection. Trans. Illum. Enging. Soc. (London), **27**, 71, 1962.
4    Lyons, S.L. Handbook of Industrial Lighting. Butterworth, London, 1981.
5    Frier, J.P. & Frier, M.E.G. Industrial Lighting Systems. McGraw-Hill, New York, 1980.

6  Boyce, P.R. & Simons, R.H. Hue discrimination and light sources. Ltg. Res. Technol., **9**, 125, 1977.
7  Commission Internationale de l'Eclairage, Publication 13.2. Method of measuring and specifying colour rendering properties of light sources. CIE, Paris, 1974.
8  British Standards Institution BS 950 Artificial daylight for the assessment of colour. BSI 1967, amended 1968.
9  National Radiological Protection Board. Protection against ultra-violet radiation in the work place, HMSO, London.
10 Gillies, G.J. Glass inspection, in Human Reliability in Quality Control, eds C.G. Drury and J.G. Fox, Taylor and Francis, London, 1975.

## APPENDIX 3   SURFACE COLOURS AND REFLECTANCES

### A3.1   The Munsell System

The colours of interior architectural surfaces are conveniently defined by the Munsell System[1], in terms of three perceptual attributes, hue, value and chroma.

Hue describes the apparently dominant part of the spectrum occupied by the colour, e.g. red, as distinct from yellow or blue. The various hues are located in different sectors of the Munsell colour solid in Fig A3.1. There are five principal hues and five intermediate hues:

principal hues: Red (R), Yellow (Y), Green (G), Blue (B), Purple (P); intermediate hues: Yellow Red (YR), Green Yellow (GY), Blue Green (BG), Purple Blue (PB), Red Purple (RP).

Chroma is the strength of the colour and increases radially in figure A3.1 from neutral grey (zero chroma) to a maximum which depends upon the hue and reflectance of the surface. Surfaces having zero chroma, and therefore no hue, are denoted Neutral (N).

Value, plotted vertically in figure A3.1, measures the lightness of the surface from 0 (perfect black) to 10 (perfect white). This is another way of describing reflectance, but differs numerically from reflectance in an important respect: each of the three Munsell scales is divided in such a way that equal intervals of hue, value or chroma denote equal steps in perceived contrast (though the value scale does not have the same spacing as the hue or chroma scales).

The relation between reflectance $R$ (per cent) and Munsell Value $V$ is given by the equation

$$R = 1.1914V - 0.22533V^2 + 0.23352V^3 - 0.020484V^4 + 0.0008194V^5$$

However, the following approximation is easily remembered, and sufficiently accurate for most practical purposes:

$$R = V(V-1)$$

The Munsell coordinates for a coloured surface comprise hue, value and chroma, in that order. For example, Munsell reference 2.5GY 6/8 indicates that the hue is 2.5GY, a distinctly yellowish green, the value is 6, which means a reflectance of about 0.3, and the chroma is 8 – moderately saturated but not startling.

Figure A3.2 shows the bounding surface of the Munsell colour solid. This indicates the relative positions of the highest possible chromas at each hue and value. Thus high chroma yellows have high value (reflectance) while high chroma reds and blues have low reflectances.

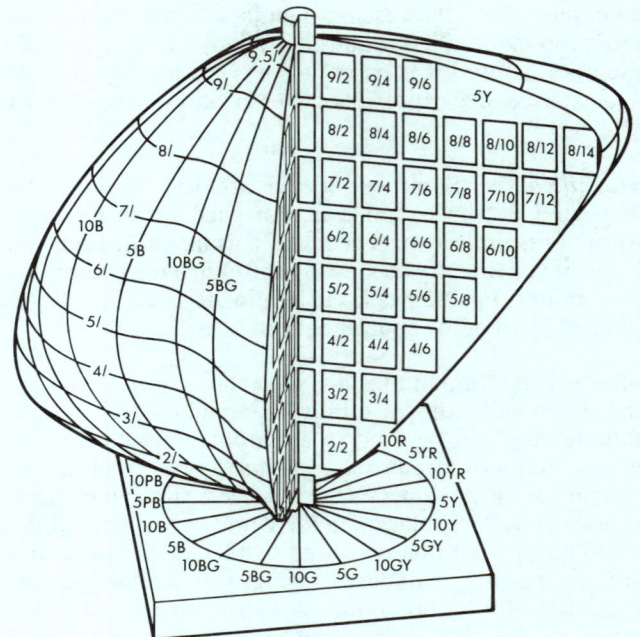

**Fig. A3.2** The Munsell Colour Solid with a slice removed showing the boundary of the 5Y (yellow) Hue plane. Produced with the kind permission of the Munsell Colour Company of Baltimore

### A3.2   British Standard BS 5252: 1976, Framework for Colour Co-ordination for Building Purposes

BS 5252 is a short-list of 237 surface colours. Though cross-referenced to Munsell coordinates the British Standard adopts another colour notation specially developed to avoid some of the anomalies of the Munsell System.

The three BS dimensions of colour are: *hue*, designated by an even number; *greyness*, designated by a letter; *weight*, designated by an additional number, usually odd.

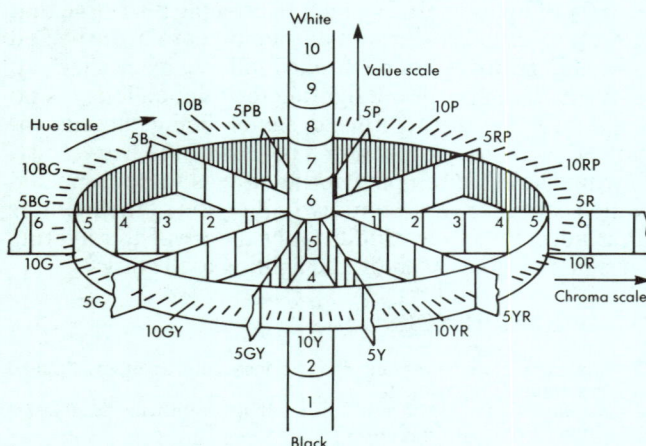

**Fig. A3.1** A schematic diagram of the layout of the Munsell Colour System. Produced with the kind permission of the Munsell Colour Company of Baltimore

Twelve distinct hues are identified, not counting 'neutral':

- 00 = neutral
- 02 = red-purple
- 04 = red
- 06 = warm orange
- 08 = cool orange
- 10 = yellow
- 12 = green-yellow
- 14 = green
- 16 = blue-green
- 18 = blue
- 20 = purple-blue
- 22 = violet
- 24 = purple

The estimated grey content of surface colours is represented on a five-point alphabetical scale:

- A = grey
- B = near grey
- C = distinct hue
- D = nearly clear
- E = clear, vivid colour

Generally greyness correlates well with chroma, give-or-take a step or two where Munsell may have gone adrift. But the two concepts are not interchangeable, for Black (oo E 53) and white (oo E 55), having no greyness content, are *clear* (Group E) not *grey* (Group A).

Surfaces having the same weight generally look equally light, but the correlation between weight and reflectance, even within a given greyness group, is uneven. In Groups A and B the correlation is excellent, in Group C it is good, but in D and E it is poorer. This is because saturated reds and blues tend to look lighter than yellows or greys of the same reflectance; the phenomenon is known as the Helmholtz-Kohlrausch effect.[3]

Individual colours are identified by a combination of a hue number, a greyness group letter and a weight number, in that order. For example, 12 B 27 means that the hue number is 12 – greenish yellow. The greyness group is B – close to grey, but with a slight hue. The weight number is 27, indicating a darkish tone. The colour is a dark olive green.

Sets of colours suitable for particular applications have been picked from BS 5252 and published as separate British Standards[4,5,6,7,8,9]. For example the BS 4800 paint colours[4] are listed in Table A3.1, with their approximate Munsell equivalents and reflectances.

## A3.3 References

[1] Munsell Colour Coporation, Munsell Book of Colour, Baltimore, USA, 1973.
[2] British Standards Institution, BS 5252, Framework for colour coordination for building purposes, BSI 1976.
[3] Padgham, C.A. and Saunders, J.E. The perception of light and colour. G. Bell and Sons, London, 1975.
[4] British Standards Institution, BS 4800, Specification for paint colours for building purposes, BSI 1981.
[5] British Standards Institution, BS 4900, Specification for vitreous enamel colours for building purposes, BSI 1976,
[6] British Standards Institution, BS 4901, Specification for plastics colours for building purposes, BSI 1976.
[7] British Standards Institution, BS 4902, Specification for sheet and tile flooring colours for building purposes, BSI 1976.
[8] British Standards Institution, BS 4903, Specification for external colours for farm buildings, BSI 1979.
[9] British Standards Institution, BS 4904, Specification for external cladding colours for building purposes, BSI 1978.

**Table A3.1 BS 5252 Colour designation, approximate Munsell reference and reflectance for colours in BS 4800, 1981**

| Greyness Group | Colour designation | Hue | Approximate Munsell reference | Approx. Reflectance |
|---|---|---|---|---|
| A | 00 A 01 | Neutral | N 8.5 | 0.68 |
| | 00 A 05 | Neutral | N 7 | 0.45 |
| | 00 A 09 | Neutral | N 5 | 0.24 |
| | 00 A 13 | Neutral | N 3 | 0.11 |
| | 10 A 03 | Yellow | 5Y 8/0.5 | 0.60 |
| | 10 A 07 | Yellow | 5Y 6/0.5 | 0.33 |
| | 10 A 11 | Yellow | 5Y 4/0.5 | 0.14 |
| B | 04 B 15 | Red | 10R 9/1 | 0.79 |
| | 04 B 17 | Red | 10R 8/2 | 0.62 |
| | 04 B 21 | Red | 10R 6/2 | 0.33 |
| | 08 B 15 | Yellow-red | 10YR 9.25/1 | 0.86 |
| | 08 B 17 | Yellow-red | 8.75YR 8/2 | 0.64 |
| | 08 B 21 | Yellow-red | 8.75YR 6/2 | 0.32 |
| | 08 B 25 | Yellow-red | 8.75YR 4/2 | 0.16 |
| | 08 B 29 | Yellow-red | 8.75YR 2/2 | 0.07 |
| | 10 B 15 | Yellow | 5Y 9.25/1 | 0.87 |
| | 10 B 17 | Yellow | 5Y 8/2 | 0.61 |
| | 10 B 21 | Yellow | 5Y 6/2 | 0.33 |
| | 10 B 25 | Yellow | 5Y 4/2 | 0.16 |
| | 10 B 29 | Yellow | 5Y 2/2 | 0.07 |
| | 12 B 15 | Green-yellow | 5GY 9/1 | 0.81 |
| | 12 B 17 | Green-yellow | 2.5GY 8/2 | 0.61 |
| | 12 B 21 | Green-yellow | 2.5GY 6/2 | 0.33 |
| | 12 B 25 | Green-yellow | 2.5GY 4/2 | 0.15 |
| | 12 B 29 | Green-yellow | 2.5GY 2/2 | 0.07 |
| | 18 B 17 | Blue | 5B 8/1 | 0.62 |
| | 18 B 21 | Blue | 5B 6/1 | 0.34 |
| | 18 B 25 | Blue | 5B 4/1 | 0.16 |
| | 18 B 29 | Blue | 7.5B 2/1 | 0.06 |
| | 22 B 15 | Violet | 10PB 9/1 | 0.81 |
| | 22 B 17 | Violet | 10PB 8/2 | 0.60 |
| C | 02 C 33 | Red-purple | 7.5RP 8/4 | 0.62 |
| | 02 C 37 | Red-purple | 7.5RP 5/6 | 0.23 |
| | 02 C 39 | Red-purple | 7.5RP 3/6 | 0.10 |
| | 02 C 40 | Red-purple | 7.5RP 2/4 | 0.07 |
| | 04 C 33 | Red | 7.5R 8/4 | 0.62 |
| | 04 C 37 | Red | 7.5R 5/6 | 0.23 |
| | 04 C 39 | Red | 7.5R 3/6 | 0.10 |
| | 06 C 33 | Yellow-Red | 7.5YR 8/4 | 0.62 |
| | 06 C 37 | Yellow-red | 5YR 5/6 | 0.23 |
| | 06 C 39 | Yellow-red | 7.5YR 3/6 | 0.11 |
| | 08 C 31 | Yellow-red | 10YR 9/2 | 0.81 |
| | 08 C 35 | Yellow-red | 10YR 7/6 | 0.46 |
| | 08 C 37 | Yellow-red | 10YR 5/6 | 0.23 |
| | 08 C 39 | Yellow-red | 10YR 3/6 | 0.10 |
| | 10 C 31 | Yellow | 5Y 9/2 | 0.81 |
| | 10 C 33 | Yellow | 5Y 8.5/4 | 0.71 |
| | 10 C 35 | Yellow | 5Y 7/6 | 0.45 |
| | 10 C 39 | Yellow | 5Y 3/4 | 0.10 |
| | 12 C 33 | Green-yellow | 2.5GY 8/4 | 0.62 |
| | 12 C 39 | Green-yellow | 2.5GY 3/4 | 0.10 |
| | 14 C 31 | Green | 5G 9/1 | 0.81 |
| | 14 C 35 | Green | 5G 7/2 | 0.45 |
| | 14 C 39 | Green | 5G 3/4 | 0.10 |
| | 14 C 40 | Green | 5G 2/2 | 0.07 |
| | 16 C 33 | Blue-green | 7.5BG 8/2 | 0.60 |
| | 16 C 37 | Blue-green | 7.5BG 5/4 | 0.22 |
| | 18 C 31 | Blue | 5B 9.25/1 | 0.84 |
| | 18 C 35 | Blue | 7.5B 7/3 | 0.42 |
| | 18 C 39 | Blue | 7.5B 3/4 | 0.10 |
| | 20 C 33 | Purple-blue | 5PB 8/4 | 0.63 |
| | 20 C 37 | Purple-blue | 5PB 5/6 | 0.23 |
| | 20 C 40 | Purple-blue | 5PB 2/4 | 0.07 |
| | 22 C 37 | Violet | 10PB 5/6 | 0.22 |
| | 24 C 33 | Purple | 7.5P 8/3 | 0.60 |
| | 24 C 39 | Purple | 7.5P 3/6 | 0.10 |

continued

| Greyness Group | Colour designation | Hue | Approximate Munsell reference | | Approx. Reflectance |
|---|---|---|---|---|---|
| D | 04 D 44 | Red | 7.5R | 4/10 | 0.16 |
| | 04 D 45 | Red | 7.5R | 3/10 | 0.10 |
| | 06 D 43 | Yellow-red | 7.5YR | 6/10 | 0.33 |
| | 06 D 45 | Yellow-red | 5YR | 4/8 | 0.16 |
| | 10 D 43 | Yellow | 5Y | 7/10 | 0.45 |
| | 10 D 45 | Yellow | 5Y | 5/10 | 0.24 |
| | 12 D 43 | Green-yellow | 2.5GY | 6/8 | 0.32 |
| | 12 D 45 | Green-yellow | 2.5GY | 4/6 | 0.15 |
| | 16 D 45 | Blue-green | 7.5BG | 3/6 | 0.10 |
| | 18 D 43 | Blue | 7.5B | 5/6 | 0.22 |
| | 20 D 45 | Purple-blue | 5PB | 3/8 | 0.10 |
| | 22 D 45 | Violet | 10PB | 3/8 | 0.10 |
| E | 04 E 49 | Red | 7.5R | 9/3 | 0.80 |
| | 04 E 51 | Red | 7.5R | 6/12 | 0.33 |
| | 04 E 53 | Red | 7.5R | 4.5/16 | 0.18 |
| | 06 E 50 | Yellow-red | 7.5YR | 8/8 | 0.60 |
| | 06 E 51 | Yellow-red | 2.5YR | 7/11 | 0.46 |
| | 06 E 56 | Yellow-red | 5YR | 5/12 | 0.24 |
| | 08 E 51 | Yellow-red | 10YR | 7.5/12 | 0.51 |
| | 10 E 49 | Yellow | 10Y | 9/4 | 0.79 |
| | 10 E 50 | Yellow | 5Y | 8.5/8 | 0.64 |
| | 10 E 53 | Yellow | 6.25Y | 8.5/13 | 0.64 |
| | 12 E 51 | Green-yellow | 2.5GY | 8/10 | 0.60 |
| | 12 E 53 | Green-yellow | 5GY | 7/11 | 0.44 |
| | 14 E 51 | Green | 2.5G | 6.5/8 | 0.34 |
| | 14 E 53 | Green | 5G | 5/10 | 0.22 |
| | 16 E 53 | Blue-green | 7.5BG | 5/8 | 0.22 |
| | 18 E 49 | Blue | 5B | 9/2 | 0.79 |
| | 18 E 50 | Blue | 7.5B | 8/4 | 0.60 |
| | 18 E 51 | Blue | 7.5B | 6/8 | 0.31 |
| | 18 E 53 | Blue | 10B | 4/10 | 0.15 |
| | 20 E 51 | Purple-blue | 5PB | 6/10 | 0.32 |
| | 00 E 53 | Black | N | 1.5 | 0.05 |
| | 00 E 55 | White | N | 9.5 | 0.85 |

test colour is expressed on a scale that gives a value of 100 for zero difference and reduces as the difference between position increases. For a single test sample colour the result of this calculation, which also includes a correction for chromatic adaptation, is called the CIE Special Colour Rendering Index.

The CIE recommends a set of test sample colours[1] which cover the Hue circle and which include some particularly important colours such as human skin. For general use a set of eight test sample colours is recommended. The average of the CIE Special Colour Rendering Indices for these eight test samples is called the CIE General Colour Rendering Index (Fig A4.1).

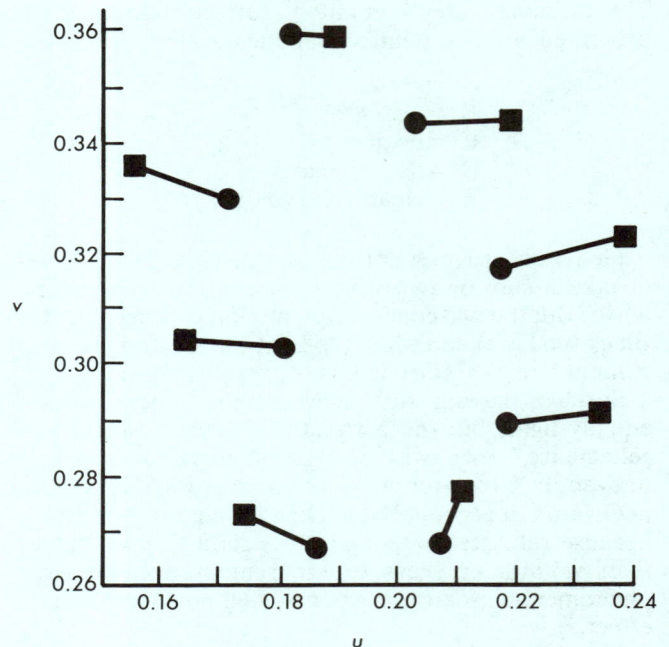

Fig. A4.1 The positions of the eight CIE test samples are shown on the CIE 1960 Uniform Chromaticity Scale diagram under the reference illuminant (squares) and under the lamp of interest (circles).

# APPENDIX 4   THE COLOUR RENDERING PROPERTIES OF LIGHT SOURCES

## A4.1   Types of colour rendering

The ability of light sources to render colours can be described by a number of different measures. Each measure characterises a particular aspect of the colour rendering properties of the light source so different measures will be appropriate in different circumstances. The measures to be considered are (a) the CIE Colour Rendering Index[1], (b) the colour gamut[2], and (c) the colour preference index[3].

## A4.2   THE CIE COLOUR RENDERING INDEX

### A4.2.1   The basis of the CIE Colour Rendering Index

By far the most widely used measure of the colour rendering properties of light sources is the CIE General Colour Rendering Index[1]. This Index quantifies the accuracy with which test sample colours are reproduced by the light source of interest relative to their colour under a reference light source. The procedure involves calculating the positions of the test sample colours on a specified uniform chromaticity scale diagram when illuminated by a test light source and a reference illuminant. The difference between the positions for a

Perfect agreement between the colours of all eight test samples under the test light source and under the reference illuminant gives a CIE General Colour Rendering Index of 100. Differences between the positions will produce lower values of the CIE General Colour Rendering Index, the values being scaled so that a warm white tubular fluorescent lamp has a CIE General Colour Rendering Index of about 50[1].

The choice of a suitable reference illuminant is obviously important and as the correction for chromatic adaptation in use at the moment is only applicable to small differences in chromaticity, a series of reference illuminants has been recommended[1]. For colour temperatures at or below 5000K, the reference illuminant is a full radiator; above 5000K it is one of a series of daylight spectral energy distributions specified by the CIE. The reference illuminant chosen has to be the one closest in chromaticity to the test light source and it is essential that the colour temperature of the reference illuminant be quoted with the CIE Colour Rendering Index.

## A4.2.2 Limitations of the CIE Colour Rendering Index

The amount of information available from any procedure that results in a single number index to describe the colour rendering properties of a light source is bound to be limited. The CIE General Colour Rendering Index indicates how the colour rendering of the test light source differs from that of the reference illuminant but it does not describe the differences in detail. To obtain this information a series of CIE Special Colour Rendering Indices is required; the greater the number of these, the more accurate the knowledge of the colour rendering properties of the test source.

The following characteristics of the CIE Colour Rendering Index should be noted. *(a)* Light sources with the same CIE General Colour Rendering Indices do not necessarily render colours in the same way because the same CIE General Colour Rendering Index may result from very different combinations of CIE Special Colour Rendering Indices. *(b)* Light sources with the same CIE Special Colour Rendering Index for a particular test colour do not necessarily render that colour in the same way. The CIE Special Colour Rendering Index does not indicate the direction of the differences in colour. *(c)* The CIE Colour Rendering Index is intended for comparing the colour rendering of two white lamps of about the same chromaticity. A lamp with a colour some distance from the full radiator locus cannot be assessed accurately. *(d)* Comparison of the colour rendering properties of light sources with different apparent colours is difficult since each CIE Colour Rendering Index refers to a reference illuminant of a chromaticity similar to that of the test light source. The colour rendering effects of illuminants with different apparent colours, e.g. daylight at a correlated colour temperature of 6500K and an incandescent lamp at a correlated colour temperature of 2700K are entirely different, although they can both have a CIE General Colour Rendering Index of 100.

## A4.3 Colour Gamut

An alternative way to describe the colour rendering properties of a light source is to plot a colour gamut[2]. To obtain the colour gamut of a light source, the colours of the eight CIE test samples used in the calculation of the CIE General Colour Rendering Index[1] are needed. The positions of these samples are plotted on the CIE 1976 Uniform Chromaticity Scale diagram[4] and joined up. The resulting octagonal figure is the colour gamut of the light source. Figure A4.2 shows the colour gamuts for a number of different light sources. It should be apparent that a single colour gamut does not indicate the accuracy with which colours are rendered. Nonetheless a great deal can be learnt from it. From the shape of the gamut and the spacing of the positions of the individual test samples, the extent to which different colours can be distinguished is apparent. From the location of the colour gamut on the CIE 1976 UCS diagram, the appearance of colours can be appreciated to some degree. The colour gamut is a convenient way of comparing the colour rendering properties of different light sources, although experience is needed in interpretation.

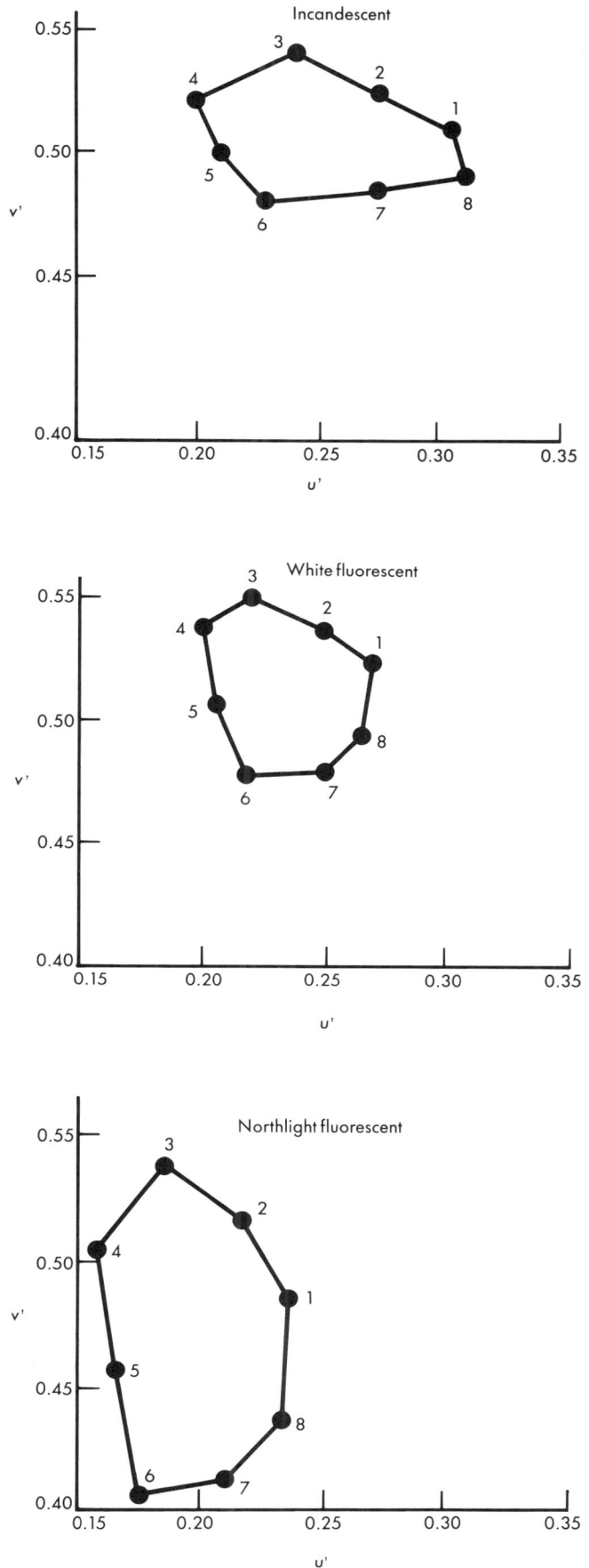

**Fig. A4.2**  Colour gamuts on the CIE 1976 Uniform Chromaticity Scale Diagram for a variety of lamps.

## A4.4 Colour Preference Index

Where good colour rendering light sources are needed for the sake of appearance, rather than precise colour matching, e.g. in a supermarket, another measure of the light source colour rendering properties is suitable. It has been established that people prefer colours to be somewhat more colourful than they are under daylight[5]. This offers the opportunity to create a Colour Preference Index based on the colour gamut for the eight CIE test samples. A colour gamut is produced for the preferred appearance of the test colours and then the colour gamut for the test samples when lit by a light source of interest is plotted on the same diagram (Fig A4.3). The closer the two colour gamuts are to each other the better the light source is at rendering the colours as preferred. A single number colour preference index based on the differences between the two gamuts has been suggested[3] but there is little doubt that more information is available from the complete colour gamuts.

## A4.5 Correlation between alternative measures

Although each of the measures discussed describes the colour rendering properties of a light source in a different way, there is considerable correlation between the order in which they rank different light sources. Light sources with a high CIE General Colour Rendering Index tend to have large colour gamuts and high colour preference indices. Conversely light sources with a low CIE General Colour Rendering Index tend to have small colour gamuts and low colour preference indices. However, the correlation between measures is not perfect and it is when the correlation breaks down that the merits of the various measures discussed need to be most carefully considered.

## A4.6 References

1. Commission Internationale de l'Eclairage. Methods of measuring and specifying colour rendering properties of light sources. CIE Publication 13.2 CIE Paris, 1974.
2. Thornton, W.A. Colour Discrimination Index. J. Opt. Soc. Amer. **62**, 191, 1972.
3. Thornton, W.A. A validation of the colour preference index. J.I.E.S., **4**, 48, 1974.
4. Commission Internationale de l'Eclairage. Official recommendations on uniform colour spaces, colour difference equations and psychometric colour terms. CIE Publication 15, Supplement 2, CIE Paris, 1978.
5. Judd, D.B. A flattery index for artificial illuminants. Illuminating Engineering, **62**, 593, 1967.

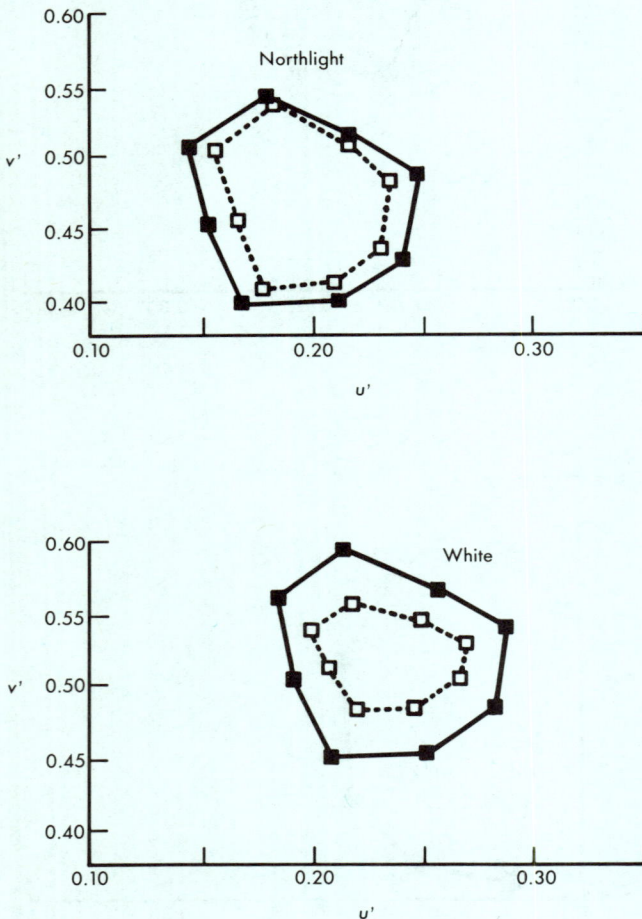

# APPENDIX 5   GLARE FORMULATIONS

## A5.1 Disability Glare

Consider a situation where a source of high luminance lies close to a task. The light from the source of high luminance will be scattered in the eye across the retinal image of the task. The effect of this scattered light is to reduce the contrast of the task and to change the local state of adaptation of the retina. Further the retinal image of the high luminance source will induce changes in the operating state of the surrounding retinal areas. The net effect of these changes is to reduce the visibility of the task. This phenomenon is known as disability glare.

The extent of disability glare can be quantified by measuring the luminance of a uniform field which, when placed over the task, produces the same reduction in task visibility as does the disability glare. This equivalent veiling luminance, as it is known, can be measured in the laboratory but for field conditions it is more convenient to calculate it. Different experimenters have produced different formulae relating the equivalent veiling luminance to the physical conditions[1,2,3] but the most generally accepted formula[3] is

$$L_v = 10 \sum_n \frac{E_n}{\theta_n^2}$$

where   $L_v$ = the equivalent veiling luminance (cd/m$^2$)

$E_n$ = the illuminance at the eye on a plane perpendicular to the line of sight, from the $n$th glare source (lux)

$\theta_n$ = the angle between the line of sight and the $n$th glare source (degrees)

**Fig. A4.3** The colour gamuts of Northlight and White Fluorescent lamps are shown (open squares) as are the positions for the preferred appearance of the eight CIE Test Samples (filled squares). The difference in position for each test sample indicates the extent to which the lamp will give the sample its preferred appearance

This formula holds for values of $\theta_n$ from 1.5° to 60°. For values of $\theta_n$ less than 1.5° a different formula has been proposed[4]. It is

$$L_v = 9.2 \sum_n \frac{E_n}{\theta_n^{3.44}}$$

where the terms have the same meaning and units as above. These two equations demonstrate that disability glare is greatest for glare sources very close to the line of sight, and diminishes rapidly with deviation from the line of sight.

The effect of equivalent veiling luminance on the contrast of a task can be estimated by adding the equivalent veiling luminance to the luminance of task detail and the luminance of the task background equally. The resulting contrast can be considered as the effective task contrast in the presence of disability glare.

The formula for effective task contrast is

$$C_e = \left| \frac{(L_t + L_v) - (L_b + L_v)}{(L_b + L_v)} \right| = \left| \frac{(L_t - L_b)}{(L_b + L_v)} \right|$$

where
$C_e$ = effective task contrast
$L_t$ = luminance of task detail (cd/m$^2$)
$L_b$ = luminance of task background (cd/m$^2$)
$L_v$ = equivalent veiling luminance (cd/m$^2$)

It should be noted that this formulation of disability glare assumes a static viewing suitation. Further, the magnitude of the disability glare is determined by the illuminance at the eye and not by the luminance of the glare source. Therefore these formulae cannot take account of situations where the luminance of the glare source is sufficient to produce after images. Fortunately, disability glare is a rare occurrence in interior lighting. Light sources which might act as sources of disability glare either tend to be located away from common lines of sight or they create sufficient discomfort to cause remedial action to be taken before significant disability glare occurs.

## A5.2 Discomfort glare from electric lighting

### A5.2.1 The Glare Index System
Discomfort glare is likely to be experienced whenever some part of the interior has a much higher luminance than occurs in the rest of the interior. By far the most common sources of high luminances in interiors are luminaires and windows. The discomfort glare effect of windows is discussed later. The magnitude of the discomfort glare which will be produced by an electric lighting installation can be estimated by calculating the glare index[5]. The basis of this index is a formula for the discomfort glare caused by small area sources, developed by Petherbridge and Hopkinson[6]. The formula for obtaining the glare index for a lighting installation is

$$\text{Glare Index} = 10 \log_{10} \frac{0.45}{L_b} \sum_n \frac{L_s^{1.6}\, \omega^{0.8}}{P^{1.6}}$$

where
$L_s$ = the luminance of the $n$th glare source (cd/m$^2$)
$L_b$ = the average luminance of the field of view, excluding the glare source (cd/m$^2$)
$\omega$ = the subtended area of the $n$th glare source (steradians)
$P$ = position index of the $n$th glare source which increases with increasing deviation from the line of sight

This formula can be applied to individual or randomly arranged or regular arrays of luminaires for any specified direction of view. A simplified tabular method based on the above formula applied to a set of standard conditions will be given in a forthcoming CIBS publication, as will be advice on computing procedures suitable for use with the basic formula[5].

The Glare Index System can be applied to a wide range of conventional luminaires. However it does have some limitations. It cannot be applied to large area light sources such as luminous ceilings because the basic formula is then invalid. It cannot be applied to coffered ceilings and similar large cut-off luminaires because of the difficulty of deciding what constitutes the luminaire. Also it may underestimate the discomfort glare for some ceiling mounted luminaires, specifically those where the luminaire luminous intensity distribution is such that the luminance of the ceiling immediately adjacent to the luminaire is greater than the luminance of the luminaire itself.

Given that a glare index has been calculated for an installation, the designer still has no indication of whether it is good or bad. Part 2 of this Code contains recommendations of limiting glare indices. These are the values of the glare index which should not be exceeded if discomfort is to be avoided. The limiting glare indices are based on formal assessments of the level of discomfort glare appropriate for different applications[7] and on field experience. The limiting glare indices are set three units apart on the Glare Index Scale, such a difference being necessary for a significant change in discomfort glare sensation to occur.

### A5.2.2 Alternative systems
Other systems for predicting discomfort glare exist. The Illuminating Engineering Society of North America has a measure called Visual Comfort Probability which indicates the percentage of people who will find the particular glare conditions acceptable[8]. This system is similar to the IES Glare Index System in that it uses a formula to predict the discomfort glare which is based on the luminance of the glare source, the background luminance, the area of the glare source and its position.

Several European countries use the European Glare Limiting Method[9]. In essence this consists of a series of curves which set limits to the luminance of the luminaires for different degrees of discomfort glare. This method is more restrictive than the IES Glare Index System and the American VCP system in that the only variable which can cause a change in the rating of glare is the luminaire luminance distribution. In addition, it can only be applied to reasonably regular arrays of luminaires. In spite of these limitations, a simplified version of this system has been adopted in the CIE Guide on Interior Lighting[10] as an interim system until an official CIE Glare System is available.

The CIE interim system applies to ordinary general lighting installations in which the luminaires are mounted overhead in a more or less regular pattern throughout the room and assumes that the effective ceiling cavity, floor cavity and wall reflectances are 0.7, 0.2 and 0.5 respectively. Luminaires are classified into those with bright sides and with an upward/downward flux fraction of 25/75, and those having dark sides and an upward/downward flux fraction of 0/100.

The CIE system uses three classes (1, 2 and 3) of 'glare control quality', and gives the luminaire average luminance at four angles in elevation measured from the downward vertical. For each class, these luminances vary according to whether the service illuminance provided by the installation is 750 lux and above or 500 lux and below, whether the luminaires have bright or dark sides, and whether they are seen broadside-on or end-on.

The relationship of the CIE interim system to the Glare Index System where the room dimension across the line of sight is not less than four times the mounting height of the luminaires above eye level, is given in table A5.1.

**Table A5.1** Comparison of glare indices

| CIE Interim System Class | Illuminance (lux) | IES Glare System |
|---|---|---|
| 1 | 750 | 17 |
| 1 | 500 | 18.5 |
| 2 | 750 | 18.5 |
| 2 | 500 | 20 |
| 3 | 750 | 21.5 |
| 3 | 500 | 23 |

The relationship between the Glare Index System and the European Glare Limiting method as used in Germany is given in a paper by Bellchambers et al[11].

### A5.2.3   The future of discomfort glare systems

As will be evident from the above, there exist a number of different systems for estimating discomfort glare. Each has its own advantages and disadvantages but all tend to rank lighting installations in a given room in the same way. Comparisons of different systems for a range of commonly occuring lighting installations show considerable agreement between the estimates of discomfort glare by the various systems[12]. It is doubtful if any one of the systems discussed can be shown to be significantly better than the others for all conditions of interest. This Code continues to use the Glare Index System because there is currently no better system and there is considerable practical experience in its use. However, it would be a considerable simplification if a single discomfort glare system could be adopted worldwide. The CIE is attempting to devise such a system based on a consensus formula of a form similar to that used in the Glare Index System[13,14]. The proposed

CIE formula for estimating discomfort glare sensation is

$$\text{CIE Glare Index} = 8\,\text{Log}_{10}2\left(\frac{(1+E_d/500)}{E_i+E_d}\sum_n \frac{L_n^2\,\omega_n}{P_n^2}\right)$$

where $L_n$ = luminance of the $n$th glare source $(\text{cd/m}^2)$

$\omega_n$ = solid angle of the $n$th glare source (steradians)

$P_n$ = a position index which takes account of the position of the $n$th glare source relative to the observer's line of sight

$E_d$ = the direct illuminance on a vertical plane at the observer's eye, from all glare sources (lx)

$E_i$ = the indirect illuminance on a vertical plane at the observer's eye, from all glare sources (lx)

This formula overcomes certain anomalies in the existing discomfort glare formulae but it remains to be seen if it ensures a significantly better fit to subjective ratings of discomfort glare sensation.

### A5.3   Discomfort from daylight

Observations of glare sensation caused by a direct view of the sky through windows have shown that the sensation of discomfort obeys different rules from that caused by small artificial light sources[15,16].

A formula to take account of the effect of a large source on the adaptation of the eye and the effects of the parts of it seen off the direct line of sight has been used to calculate the effects of different window sizes and sky luminances. These effects have been related to the observed discomfort and lead to the conclusion that, above a certain very small size of window, the discomfort glare produced is practically independent of window size and of distance from the observer. A small change can be made by altering the average reflectance of the interior room surfaces, but by far the greatest effect is that of the luminance of the sky.

The frequency of occurrence of various values of average luminance of the sky can roughly be inferred from available data on the frequency of occurrence of the corresponding values of diffuse sky illuminance. These data suggest that for the U.K., an unprotected window will be uncomfortably glaring over a significant period of the year. The design of the window openings and their screening are therefore important in the avoidance of discomfort glare in buildings during the daytime.

Usually, the most effective precaution is to provide adjustable screening in the form of light curtains, blinds or louvres. Window design to reduce the contrast of the areas immediately surrounding the bright sky also plays a part in reducing discomfort. (See section 4.4.1.4.)

## A5.4 References

1. Stiles, W.S. The effect of glare on the brightness difference threshold. Proc. Roy. Soc. B104, 322, 1929.
2. Holladay, L.L. The fundamentals of glare and visibility. J. Opt. Soc. Amer. **12,** 271, 1926.
3. Commission Internationale de l'Eclairage, CIE Publication 31, Glare and uniformity in road lighting installations, CIE, Paris, 1976.
4. Hills, B.L. Visibility under night driving conditions: derivation of (L,A) characteristics and factors in their application. Ltg. Res. & Technol. **8,** 11, 1976.
5. Chartered Institution of Building Services. The Calculation of Glare Indices, to be published.
6. Petherbridge, P. & Hopkinson, R.G. Discomfort Glare and the lighting of buildings. Trans. Illum. Engng. Soc. (London) **15,** 39, 1950.
7. Luminance Study Panel of the IES Technical Committee. The development of the IES Glare Index System. Trans. Illum. Engng. Soc. (London), **27,** 9, 1962.
8. Illuminating Engineering Society of North America. Lighting Handbook, 1981.
9. Fischer, D. The European Glare Limiting Method. Ltg. Res. & Technol. **4,** 97, 1972.
10. Commission Internationale de l'Eclairage. Guide on Interior Lighting. CIE Publication 29, CIE Paris 1975.
11. Bellchambers, H.E., Collins, J.B. & Crisp, V.H.C. Relationship between two systems of glare limitation. Ltg. Res. & Technol. **7,** 106, 1975.
12. Manabe, H. The assessment of discomfort glare in practical lighting installations. Oteman Economic Studies (9). Oteman Gakuin University, Osaka, Japan 1976..
13. Commission International de l'Eclairage, TC 3.4, Discomfort Glare, Committee Report on methods of estimating discomfort glare, CIE Paris 1983.
14. Lowson, J.C. Practical application of the Einhorn (CIE) Glare Index formula. Ltg. Res. and Technol., **13,** 169, 1981.
15. Hopkinson, R.G., Longmore, J., Woods, P.C. & Craddock, J. Glare discomfort from windows, CIE 17th Session Barcelona, 1971.
16. Chauvel, P., Collins, J.B., Dogniaux, R. and Longmore, J. Glare from windows: current views of the problem. Ltg. Res. & Technol., **14,** 31, 1982.

## APPENDIX 6  CONTRAST RENDERING FACTOR

### A6.1  Definition of Contrast Rendering Factor

The luminance contrast of a target is defined as

$$\text{Luminance contrast}, C = \left| \frac{L_t - L_b}{L_b} \right|$$

where $L_t$ = luminance of the target (cd/m$^2$)
$L_b$ = luminance of the background (cd/m$^2$)

The vertical lines are a modulus symbol which means that the sign of the contrast is ignored.

The Contrast Rendering Factor is most conveniently defined[1] as a ratio of luminance contrasts for a given target. Specifically, the Contrast Rendering Factor of a target occurring at a given position under a particular lighting installation and viewed from a stated direction is defined as:

$$CRF = C_1/C_2$$

where $CRF$ = Contrast Rendering Factor

$C_1$ = luminance contrast of the target under the lighting of interest

$C_2$ = luminance contrast of the target under reference lighting viewed from the same direction

The reference lighting conditions are completely diffuse and unpolarised illumination such as can be produced in a photometric integrating sphere[1]. The reference lighting condition is chosen for the ease with which it can be produced rather than its inherent quality.

Contrast Rendering Factor can vary from zero to greater than unity. When $CRF$ approaches zero the contrast of the target under the lighting of interest is greatly reduced from what it is under reference lighting conditions. When $CRF$ is greater than unity the contrast of the target is greater under the lighting of interest than it is under the reference lighting. When $CRF$ is equal to unity the contrast of the target is the same under the lighting of interest and under reference lighting conditions.

### A6.2  Features which affect Contrast Rendering Factor

A Contrast Rendering Factor is specific to a particular target at a particular point under a particular lighting installation, viewed from a particular direction. Changing any one of these features of the situation may change the $CRF$. Thus there is no such thing as a $CRF$ for a particular lighting installation, there will always be a range of $CRF$ values[3].

To estimate what this range is likely to be in any situation it is necessary to understand the factors which determine the $CRF$ value. These factors have a common link in that they are related to the occurrence of veiling reflections. Figure A6.1 shows an example of veiling reflections. It consists of two samples one of matt print on matt paper, the other of pencil writing on matt paper. Both samples are lit in the way described in the figure. There are very few veiling reflections in the matt print on matt paper for either method of lighting but for the pencil writing on paper, when the lighting is above and in front of the observer, the light source is reflected towards the viewer with the consequence that strong veiling reflections occur. This means that the contrast of the writing on the paper is reduced. It will be appreciated that the $CRF$ of the pencil writing on matt paper will be much less than for the matt print on matt paper for the same lighting conditions.

This observation implies that an important factor in determining $CRF$ is the specularity of the reflection properties of the target; the greater the specularity the larger the range of $CRF$ values that will occur in an interior.

Another factor which is important is the geometry of the lighting/target/viewer assembly. When a source of high luminance (usually a light source) is positioned such that it is specularly reflected towards the viewer, veiling reflections can occur. The position from which the light which is specularly reflected from the target to the viewer originates is called the offending zone (Fig A6.2). It is only when a source of relatively high luminance occupies the offending zone that veiling reflections occur. For reference lighting conditions the offending zone has the same luminance as all the other parts of the photometric sphere so few veiling reflections will occur, but when a conventional ceiling mounted luminaire occupies the offending zone it will cause veiling reflections because it has a much higher luminance than most other parts of the room. This implies that the other important factor determining the range of $CRF$ values that occur in an interior is the relative luminance of the

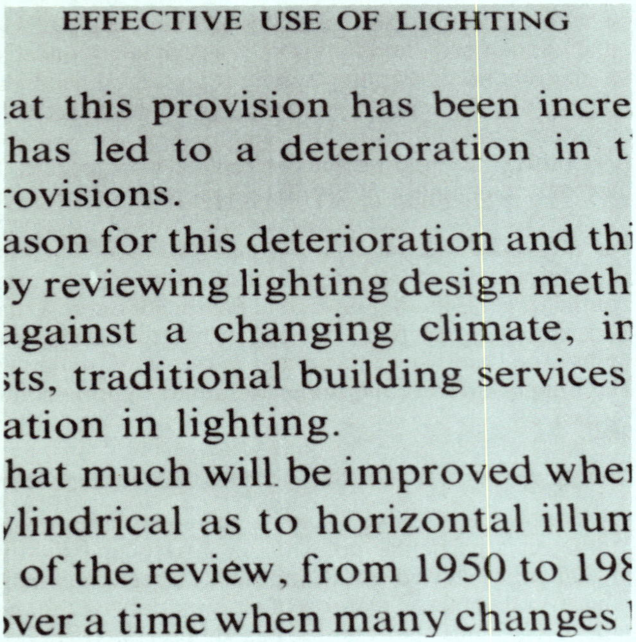

## EFFECTIVE USE OF LIGHTING

at this provision has been incre
has led to a deterioration in t
rovisions.
ason for this deterioration and thi
y reviewing lighting design meth
against a changing climate, in
sts, traditional building services
ation in lighting.
hat much will be improved whei
ylindrical as to horizontal illum
of the review, from 1950 to 198
ver a time when many changes

a) Matt ink on matt paper lit from above and behind the observer

this provision has t
hich has led to c
dustrial lighting pri
for this deteriora
n explanation by
sent and judging
ed by market pre
industry procedures,

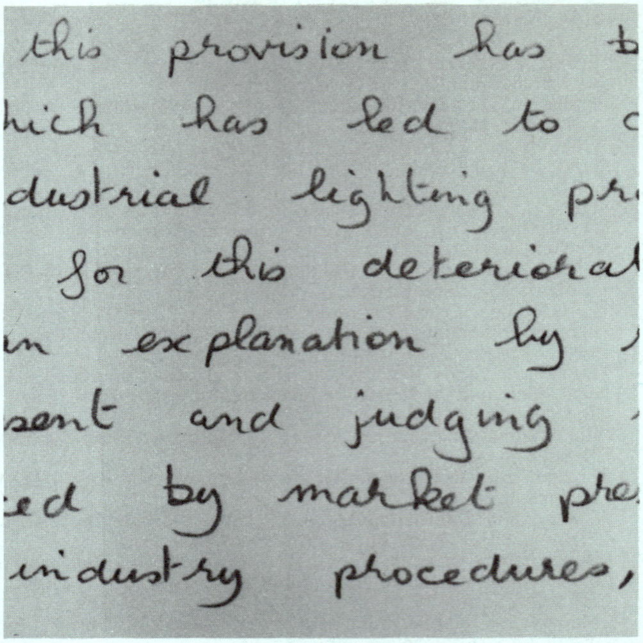

b) Pencil writing on matt paper lit from above and behind the observer

## EFFECTIVE USE OF LIGHTING

at this provision has been incr
has led to a deterioration in
rovisions.
ason for this deterioration and tl
y reviewing lighting design metl
against a changing climate, i
ts, traditional building service
ation in lighting.
hat much will be improved whe
lindrical as to horizontal illur
of the review, from 1950 to 19
ver a time when many changes

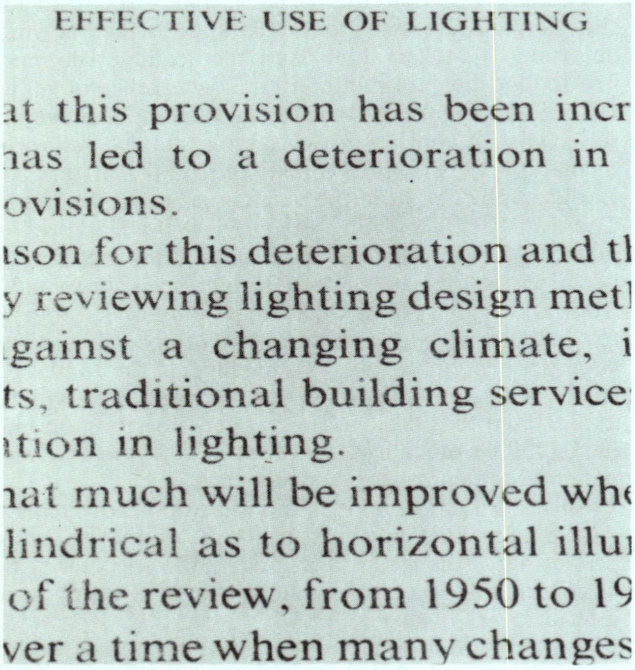

c) Matt ink on matt paper lit from above and in front of the observer

this provision has t
hich has led to c
dustrial lighting pr
for this deteriora
an explanation by
sent and judging
ed by market pl
industry procedures,

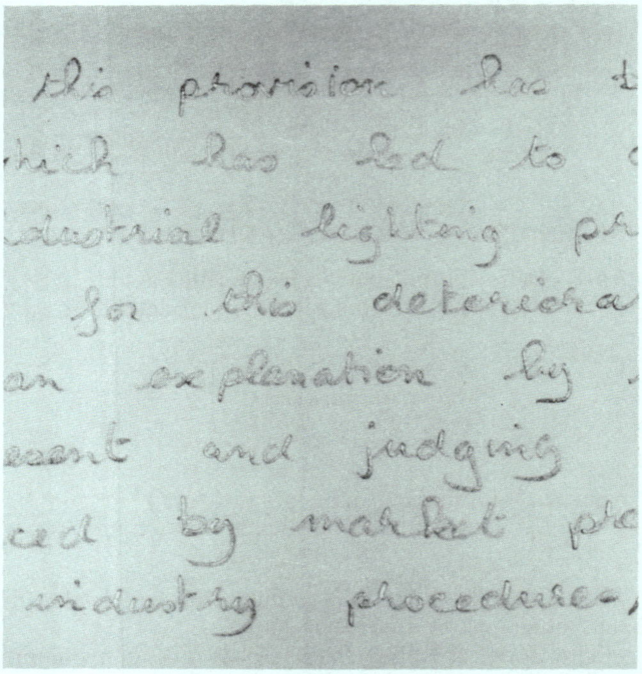

d) Pencil writing on matt paper lit from above and in front of the observer

**Fig. A6.1** Differences in specular reflections and hence in visibility for two different lighting arrangements and two different tasks

offending zone. In practical terms this usually means that the important factors are the position and viewing direction of the target, and the luminance of the parts of the luminaires which may occupy the corresponding offending zones.

The practical implications of these aspects are discussed in reference 3. However, the conclusions can be summarised in two general rules: *(1)* the more specular the target material, the greater will be the range of *CRF* values in an interior; *(2)* the smaller and more concen-

trating the luminaires, the greater will be the variation of *CRF* with position. These two rules act in combination. Very large variations in *CRF* will occur with position for specular targets in a room lit by very concentrating luminaires. Very little variation in *CRF* will occur with positions for matt targets in a room lit by very diffuse lighting, e.g. a luminous ceiling.

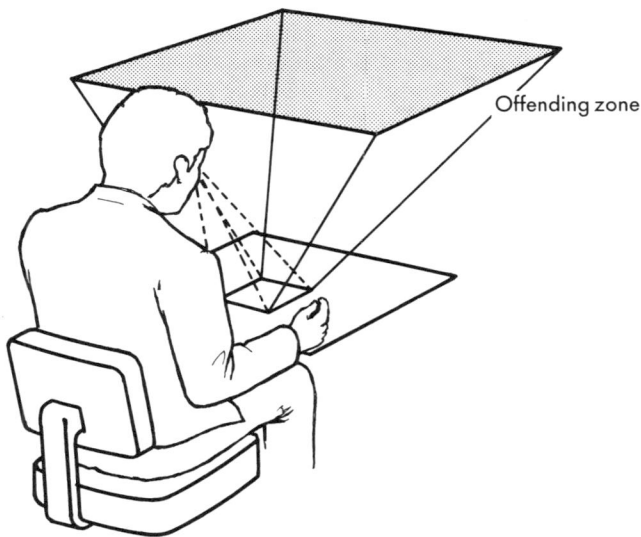

**Fig. A6.2** A schematic diagram of the 'offending zone'

## A6.3 Effects of low Contrast Rendering Factor values

Low CRF values can affect both visual performance and visual comfort. Because low CRF values imply reduced contrast, where task performance is dependent on contrast then performance will be reduced. In practice this occurs rarely, because for most situations either the reduction in contrast is not important or the CRF value can be restored to a higher value by changing the viewing position. However, where the task contrast is critical to the performance of the task, then care should be taken to avoid low CRF values.

The most usual effect of low CRF values is that of visual discomfort. Even when performance can be maintained in the presence of veiling reflections, the knowledge that the lighting is reducing the contrast of the task is irritating and will give rise to complaints about the lighting. Figure A6.3 shows some results of people's assessments of the degree of disturbance felt with different materials seen at different levels of CRF[4]. It can be seen that as CRF values decrease the level of disturbance increases. From these and other results[5,6,7] it can be concluded that a CRF value of 0.7 is the minimum that should occur at work stations if visual discomfort is to be avoided. It should be noted that this criterion does not apply to very glossy materials. For such materials it is possible to have a state such that contrast reversal occurs, i.e. glossy black lettering becomes brighter than matt white paper. Contrast reversal will give a high CRF value but it is considered an unsatisfactory aspect of the lighting[7].

## A6.4 Prediction of Contrast Rendering Factor

When designing a new installation, the Contrast Rendering Factor has to be calculated rather than measured. Completely general calculations which can be applied to any target and any installation are not easy. They require computer programs of some complexity, considerable computer time and detailed information about the photometric performance of the luminaire and the reflection characteristics of the target. Such programs exist[8] and the necessary photometric information is available for a few targets. The principles used in the calculation procedure are discussed in reference 9.

However, a more limited but more easily applied procedure has been developed[10]. This calculation procedure, which only applies to a pencil target on a matt background, results in a diagram analogous to the isolux diagram (see section 4.5.4) which specifies the CRF at various positions relative to a single luminaire or, for a regular array of luminaires, in an average Contrast Rendering Factor. It is expected that this latter result will allow manufacturers to produce tables of average CRF for their luminaires, similar to the existing utilisation factor tables. With this information, and a check on the minimum CRF that is likely to occur, using the isolux diagram approach, it should be possible for the designer to predict the levels of CRF that are likely to occur for pencil writing on a matt background.

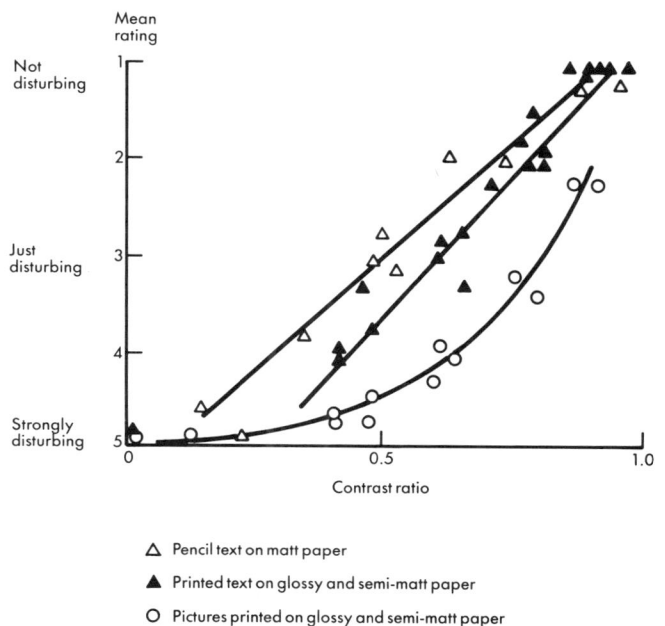

△ Pencil text on matt paper

▲ Printed text on glossy and semi-matt paper

O Pictures printed on glossy and semi-matt paper

**Fig. A6.3** Mean ratings of disturbance caused by veiling reflections for various materials and contrast rendering factors

## A6.5 Measurement of Contrast Rendering Factor

### A6.5.1 Exact measurements

For an existing installation the Contrast Rendering Factor at different positions can be measured. Exact measurements of Contrast Rendering Factor require accurate measurements of target and background luminance to be made under the actual and reference lighting conditions. From these luminances the luminance contrast of the target under actual and reference lighting conditions can be calculated and hence the CRF can be obtained. An instrument for doing this rapidly is commercially available but it uses a standard target. For any target the luminance measurements can be made with a luminance meter provided it has a field of measurement appropriate to the target[3,8].

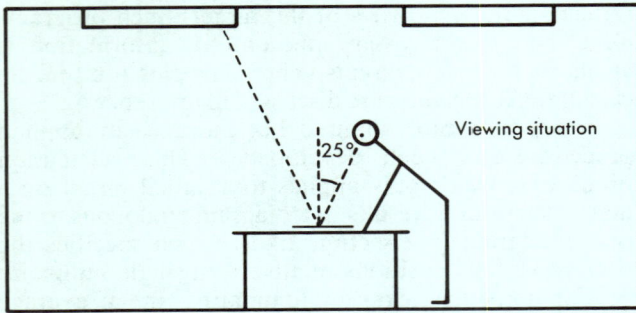

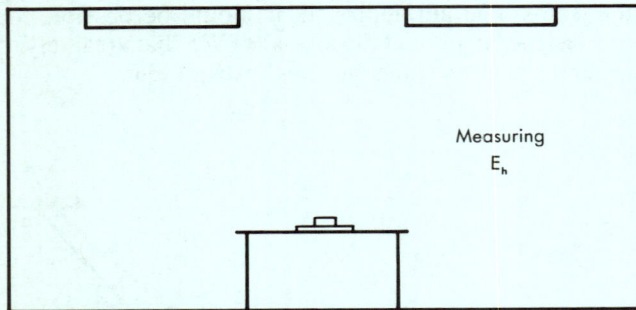

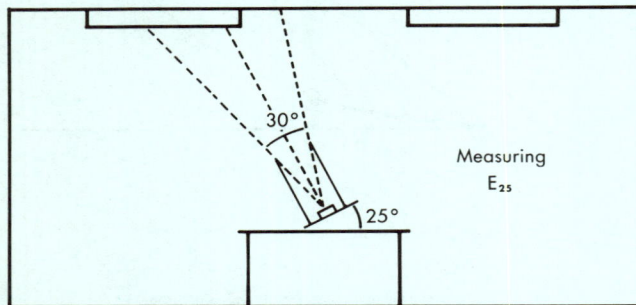

Fig. A6.4  Measurement methods for $E_h$ and $E_{25}$

### A6.5.2  Approximate measurements

If Contrast Rendering Factor is to be used simply as a criterion of visual discomfort then an approximate measurement is usually sufficient. For approximate measurements two techniques are available. The first uses an illuminance meter[3]. With this meter, illuminance measurements are taken on the horizontal plane $E_H$ and on a plane inclined at 25° from the horizontal $E_{25}$. For the inclined plane illuminance, the acceptance angle of the illuminance meter is restricted to ± 15° about the normal to the photocell (Fig A6.4). Then the factor $W$ is calculated where

$$W = \frac{E_H - E_{25}}{E_H}$$

and applied to a calibration graph which gives the corresponding $CRF$ values that will be found for pencil, ballpoint pen and dry transfer printing on matt paper (Fig A6.5). This method uses simple equipment but is

restricted to a few target materials and one viewing angle.

The second method uses a gauge consisting of an array of pencil drawn circles on matt backgrounds of different reflectance (Fig A6.6)[10]. By moving the array over the position of interest the circle which disappears can be identified. This establishes the $CRF$ value of the target for which the gauge is calibrated, for the position of interest. The gauges currently available are calibrated for pencil writing on a white matt background. This method uses very simple apparatus and is easy to use.

### A6.5.3  Indentifying positions for measurement

As $CRF$ can vary widely with position in an interior, for the purposes of checking if the 0.7 criterion has been reached it is useful to be able to identify the positions where $CRF$ is likely to be close to the minimum. This can be done by first identifying the sources of high luminance in the room, and then establishing the positions of interest for which these sources occupy the offending zone. These positions will have the minimum $CRF$ values, although what that minimum value is will depend on the specularity of the material.

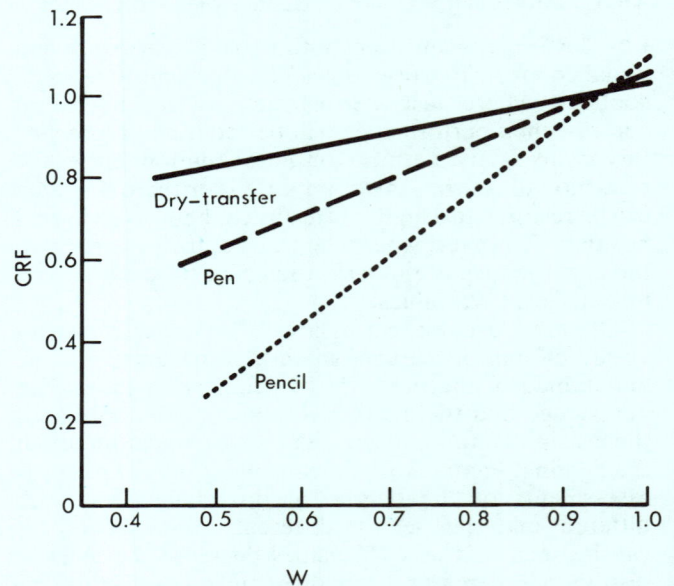

Fig. A6.5  Regression lines for predicting CRF values for pencil, pen or dry-transfer material on white matt paper when seen from 25° from the vertical

### A6.6  References

1  Commission Internationale de l'Eclairage. An analytical method for describing the influence of lighting parameters upon visual performance. CIE Publication 19/2, CIE Paris 1981.
2  Boyce, P.R. The variability of Contrast Rendering Factor in lighting installations. Ltg. Res. & Technol. 10, 94-105, 1978.
3  Boyce, P.R. & Slater, A.I. The application of Contrast Rendering Factor to office lighting design. Ltg. Res. & Technol. 13, 65-79, 1981.
4  De Boer, J.B. Performance and comfort in the presence of veiling reflections. Ltg. Res. & Technol. 9, 169-176, 1977.
5  Bjorset, H.H. & Frederiksen, E. A proposal for recommendations for the limitation of contrast reduction in office lighting. Proceedings CIE, 19th Session, Kyoto, 1979.
6  Boyce, P.R. Veiling reflections: an experimental study of their effects on office work, Electricity Council Research Centre, M1230, 1979.

[7] Reitmieir, J. Some effects of veiling reflections in papers. Ltg. Res. & Technol. **11**, 204-209, 1979.
[8] Slater, A.I. Variation and use of Contrast Rendering Factor and equivalent sphere illumination. Ltg. Res. & Technol. **11**, 117-139, 1979.
[9] Illuminating Engineering Society of North America. IES Lighting Handbook, 6th Edition, Reference Volume, 9-63, 1981.
[10] Lynes, J.A. Designing for contrast rendition. Ltg. Res. & Technol. **14**, 1, 1982.

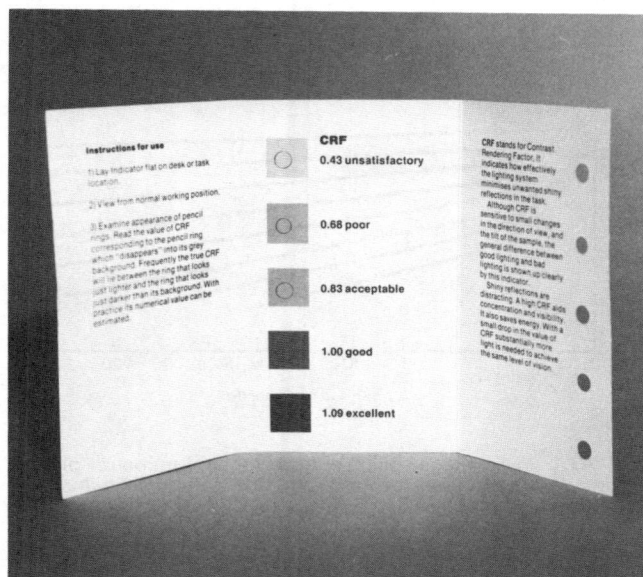

**Fig. A6.6** A gauge used to determine the CRF value at a given position for pencil writing on a white matt background

## APPENDIX 7  LIGHTING MAINTENANCE SCHEDULES AND LIGHT LOSS FACTOR

### A7.1  Lighting Maintenance Schedules

Electric lighting installations need to be maintained if they are to operate efficiently. Therefore a maintenance schedule is an essential part of any lighting design. The maintenance schedule which is considered appropriate for any particular installation is largely determined by the circumstances. The important factors are the rate at which the performance of the installation deteriorates, the consequences of such deterioration for the activities in the interior; the cost of the labour and equipment used in maintaining the installation; the effect of carrying out lighting maintenance on the activities in the interior; and the importance attached to the appearance of the installation. On occasions one of these factors will dominate the others, e.g. in a factory, maintenance may only be possible when the production process is shut down, but in general the designer has to balance the various factors to achieve an appropriate schedule. Most of the factors mentioned above are matters for managerial judgement and are beyond the scope of this Code. However, one that is within the scope of this Code is the rate at which the performance of the installation deteriorates. This can be quantified by the Light Loss Factor.

### A7.2  Light Loss Factor

The Light Loss Factor *(LLF)* is the ratio of the illuminance produced by the lighting installation at some specified time to the illuminance produced by the same installation when new. It estimates the effect on the illuminance provided by the installation of the fall in lamp luminous flux with hours of use, and the depreciation caused by dirt deposited on luminaires and room surfaces.

### A7.3  Determination of Light Loss Factor

Light Loss Factor is the product of three other factors:

Light Loss Factor = Lamp Lumen Maintenance Factor
× Luminaire Maintenance Factor
× Room Surface Maintenance Factor.

Lamp Lumen Maintenance Factor estimates the decline in light output of the light source over a specified time. Luminaire Maintenance Factor estimates the effect of dirt deposited on or in the luminaire over a set time on the light output of the luminaire. Room Surface Maintenance Factor estimates the effect of dirt deposited on the room surfaces over a set time on the illuminance produced by the installation.

#### A7.3.1  Lamp Lumen Maintenance Factor

The lumen output from nearly all lamp types reduces with time. The reduction in light output that occurs with time varies for different lamp types and it is essential to consult the manufacturer's data. From such data it is possible to obtain the Lamp Lumen Maintenance Factor for a specific number of hours of operation; the Lamp Lumen Maintenance Factor being the proportion of the initial light output that is produced after a specified time. It should be noted that manufacturers' data will be based on assumptions about the ambient temperatures in which the lamp will operate, the voltage supplied to the lamp and the control gear which is used. If any of these aspects of the proposed design are unusual then the manufacturer should be informed when the data is requested, because such differences influence the light output of the lamp over time.

#### A7.3.2  Luminaire Maintenance Factor

Dirt deposited on or in the luminaire will cause a reduction in light output from the luminaire. The rate at which dirt is deposited on or in the luminaire depends on the construction of the luminaire and on the extent to which dirt is present in the atmosphere, which in turn is related to the amount and nature of the dirt generated in the space and admitted from the surrounding area. Figure A7.1 shows typical changes in light output from a luminaire cause by dirt deposition, for a number of luminaire/activity/location categories. The categories used are defined in table A7.1. Each category is identified by a luminaire type, a place where the activity occurs, and a number of possible locations for the place. The luminaires cover a range from bare battens through sealed luminaires to indirect cornice lighting. The places are classified into three on the basis of the amount of dirt generated by the activities. The location is classified into five classes, depending on the amount of dirt likely to be present and the extent to which the air is filtered before entering the place.

From table A7.1 it is possible to identify the appropriate luminaire/activity/location category for most circumstances and hence to establish in Fig A7.1 the reduction in light output from the luminaire due to dirt deposition. From Fig A7.1 the Luminaire Maintenance Factor for a particular time can be established, the Luminaire Maintenance Factor being the proportion of the initial light output from the luminaire that occurs after the set time. For example, for a luminaire/activity/location category C the luminaire maintenance factor after 12 months is 0.77.

### A7.3.3 Room Surface Maintenance Factor

Changes in room surface reflectance caused by dirt deposition will cause changes in the illuminance produced by the lighting installation. The magnitude of these changes are governed by the extent of dirt deposition and the importance of inter-reflection to the illuminance produced. The importance of inter-reflection is closely related to the distribution of light from the luminaire. For luminaires which have a strongly downward distribution, i.e. direct luminaires, inter-reflection has little effect on the illuminance produced on the conventional horizontal working plane. Conversely, indirect lighting is completely dependent on inter-reflections. Most luminaires lie somewhere between these extremes so most lighting installations are dependent to some extent on inter-reflection. Figure A7.2 shows the typical changes in the illuminance from an installation that occur with time due to dirt deposition on the room surfaces for very clean, average and very dirty rooms lit by direct, semi-direct and indirect luminaires. Very clean rooms are typically offices, laboratories, etc. Rooms of average cleanliness occur in light industry. Very dirty rooms occur in foundries, rubber processing plants, etc. From Fig A7.2 it is possible to select a Room Surface Maintenance Factor appropriate to the circumstances, the Room Surface Maintenance Factor being the proportion of the initial illuminance produced by the installation that occurs after the specified time. For example, for a semi-direct luminaire in a room of average cleanliness after 2 years of use, the room surface maintenance factor will be 0.85.

### A7.3.4 Example of Light Loss Factor Calculation

It is desired to calculate the Light Loss Factor (*LLF*) for a bare batten tubular fluorescent installation to be used in a light industrial assembly shop on the outskirts of a town. The light loss factor is to be determined after the luminaires have been installed for 12 months. During that time the luminaires will have been operating for 2500 hours. The calculation process is as follows:

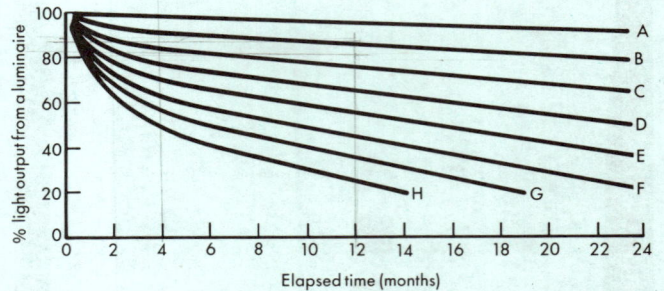

**Fig. A7.1** Percentage light output for a luminaire plotted against elapsed time for different luminaire/activity/location categories (See Table A7.1)

*From manufacturer's data*

Lamp Lumen Maintenance Factor (*LLMF*) for tubular fluorescents after 2500 hours = 0.92.
*From table A7.1*

Luminaire/activity/location category for bare battens in light industrial premises on town outskirts = B/C.
*From fig A7.1*

Luminaire Maintenance Factor (*LMF*) after 12 months for category B/C = 0.85.
*From fig A7.2*

Room Surface Maintenance Factor (*RSMF*) after 12 months for semi-direct luminaires in a room of average cleanliness = 0.88.

**Table A7.1** Luminaire/activity/location categories

| Activity | Location | Bare lamp batten | Open ventilated reflector | Dust tight, dust proof or reflector lamp | Open non-ventilated reflector, enclosed diffuser/controller | Open base diffuser or louver | Recessed diffuser or louver, diffusing or louvered luminous ceiling | Indirect cornice |
|---|---|---|---|---|---|---|---|---|
| Offices, shops & stores, hospitals, laboratories, schools, etc. | All air-conditioned buildings | A | A | A | A/B | A/B | A | B |
| | Clean country area | A/B | A/B | A/B | B | B | A/B | C/D |
| | City or town outskirts | B | B | B | C | B/C | B | E |
| | City or town centre | B/C | B/C | B/C | C/D | C | B/C | F/G |
| | Dirty industrial area | C | C | B/C | D | C D | C | G |
| Manufacturing areas, machine shops, etc. | All air-conditioned buildings | A/B | A | A | C | B/C | B | B/C |
| | Clean country area | B | A/B | B | C/D | C | B/C | D/E |
| | City or town outskirts | B/C | B | B | D | C/D | C | F |
| | City or town centre | C | B/C | B/C | D/E | D | C/D | G |
| | Dirty industrial area | C/D | C | C | E | D/E | D | H |
| Steelworks, foundries, welding shops, etc. | Clean country area | C | B/C | B | D/E | D | C/D | – |
| | City or town outskirts | C/D | C | B/C | E | D/E | D | – |
| | City or town centre | D | C/D | B/C | E/F | E | D/E | – |
| | Dirty industrial area | D/E | D | C | F | E/F | E | – |

Now,

$$LLF = LLMF.LMF.RSMF$$

*Therefore*

$$LLF = 0.92 \times 0.85 \times 0.88 = 0.69$$

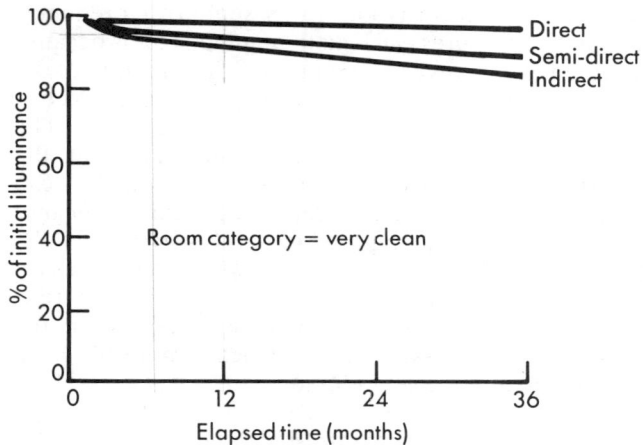

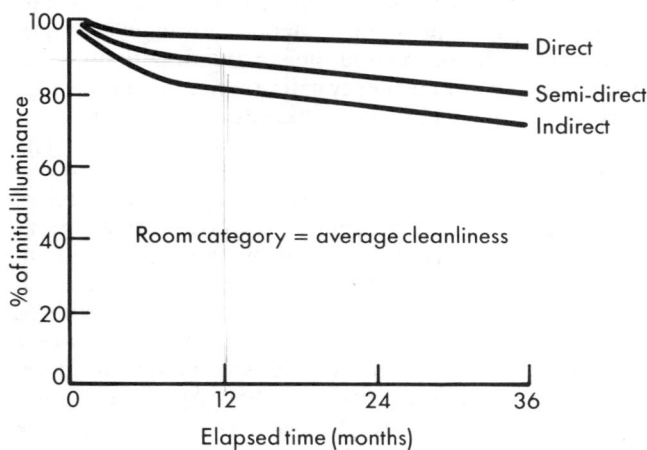

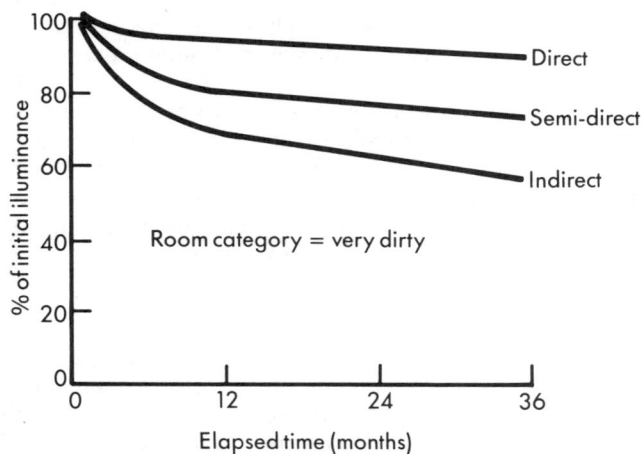

**Fig. A7.2** Percentage of initial illuminance from an installation plotted against elapsed time, for luminaires of different light distribution in rooms of different categories of cleanliness

## A7.4 Use of Light Loss Factor

Light Loss Factor can be used in the lumen method of illuminance calculation to estimate what the illuminance produced by the installation will be at any particular stage in its life. This is achieved by using the following formula:

$$E = \frac{F_{in}.n.N.UF.LLF}{A}$$

where
$E$ = illuminance (lux)
$F_{in}$ = initial luminous flux of the light source (lumen)
$n$ = the number of lamps per luminaire
$N$ = the number of luminaires
$A$ = area to be lit (metre$^2$)
$UF$ = Utilisation Factor for the luminaire in the room
$LLF$ = Light Loss Factor

By calculating the Light Loss Factor for different times and taking into account the proposed maintenance schedule, it is possible to predict the pattern of illuminance that will be produced by the installation over time (Fig 3.1). This patten can be used to assess the suitability of any proposed maintenance schedule. It can also be used to estimate the average illuminance provided by the installation over time, and hence to determine whether the installation is likely to meet the appropriate design service illuminance criterion recommended in Part 2 of this Code.

## A7.5 Light Loss Factor and Maintenance Factor

In previous Codes the effect of hours of operation on the light output from lamps has been allowed for by using Lighting Design Lumens rather than initial lumens and the effect of dirt depreciation has been allowed for by using Maintenance Factor. The Maintenance Factor is the ratio of the illuminance provided when the installation is in an average condition of dirtiness to the illuminance provided when it is clean. The equation for calculating illuminance using this approach is

$$E = \frac{F.n.N.UF.MF}{A}$$

where
$E$ = illuminance (lux)
$F$ = luminous flux of the light source at 2000 hours (in lumens), i.e. the lighting design lumens
$n$ = number of lamps per luminaire
$N$ = number of luminaires
$A$ = area to be lit (in square metres)
$UF$ = Utilisation Factor of the luminaire in the room
$MF$ = Maintenance Factor

This system has been abandoned because it gives only a single estimate of the illuminance that will be provided by the installation. Dealing with maintenance by means of the Light Loss Factor allows for a more comprehensive examination of the effect of the operating conditions and maintenance procedures on the illuminance provided.

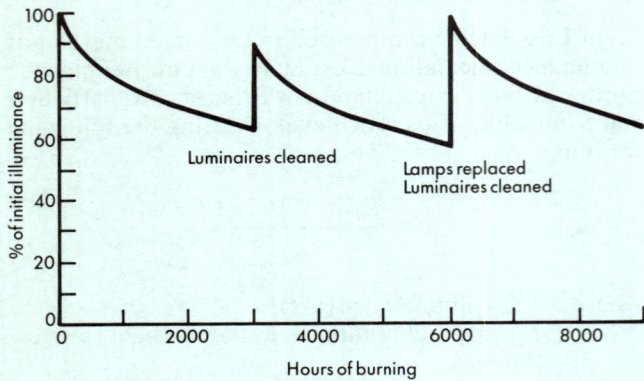

**Fig. A7.3** Variation in illuminance provided by an installation over a maintenance cycle

## APPENDIX 8   PHYSIOLOGICAL EFFECTS OF OPTICAL RADIATION

### A8.1   Introduction

In physical terms, light is a small part of the electro-magnetic spectrum, lying in the wavelength range 400 nm to 780 nm (Fig A8.1). Adjacent to it are ultra-violet radiation, nominally occupying the wavelength range 100 nm to 400 nm and infra-red radiation, covering the wavelength range 780 nm to 1 mm. The electro-magnetic radiation in the complete wavelength range 100 nm to 1 mm is called optical radiation. One effect of the part of optical radiation called light is to allow the human visual system to operate, but the non-visual part of optical radiation can also affect human eye and skin tissue. This appendix summarises the effects of optical radiation on skin and eye tissue and lists the most widely recognised criteria by which the effects of such radiation may be assessed.

### A8.2   The effects of optical radiation

#### A8.2.1   Ultra-violet radiation

The Commission Internationale de l'Eclairage has divided the ultra-violet region of the electro-magnetic spectrum into three parts: UV-A (400-315 nm), UV-B (315-280 nm) and UV-C (280-100 nm)[1]. The main hazard to human tissue comes from UV-B and UV-C radiation. Radiation from these two regions of the ultra-violet spectrum is absorbed by the cornea and conjunctiva of the eye (Fig A8.2) and in sufficient quantities will cause kerato-conjunctivitis. This is an unpleasant but temporary condition involving photo-conjunctivitis and an inflammation of the cornea called photokeratitis. The action spectrum for photo-conjunctivitis peaks at about 260 nm whilst that for photo-keratitis has a peak sensitivity at about 270 nm [2]. It should be appreciated that the important factor in the incidence of kerato-conjunctivitis is the dose of radiation received, i.e. the product of the irradiance received and the time for which it is received. Therefore both short duration, high irradiance exposures and long duration, low irradiance exposures can be harmful.

Kerato-conjunctivitis occurs at the eye, but ultra-violet radiation can also affect the skin. Exposure of the skin to ultra-violet radiation produces two distinct effects depending on the wavelength range of the incident radiation. If the skin is exposed to UV-A radiation immediate pigment darkening occurs without any reddening of the skin. This change in pigment is of short duration[2]. If the skin is exposed to UV-B and UV-C radiation reddening of the skin occurs some hours after exposure. This reddening is called erythema, and in an extreme form is known as sunburn[2]. Frequent, moderate exposures can lead to tanning. This is because repeated exposures cause changes in the structure of the skin. A new darker pigment is formed near the surface of the skin and the outer layer of the skin thickens. The effect of these changes in skin structure is to reduce the absorption of ultra-violet radiation by the skin. This can legitimately be called a defence mechanism because chronic exposure to high levels of UV-B and UV-C radiation accelerates skin ageing and increases the risk of developing certain types of skin cancer[2,3]. There is little agreement on the wavelength sensitivity for the various forms of skin cancer which can result from exposure to ultra-violet radiation.

#### A8.2.2   Visible and near infra-red radiation

Radiation in the visible and near infra-red radiation of the electro-magnetic spectrum (400-1400 nm) is transmitted through the ocular media of the eye and is absorbed by the retina and the surrounding tissues (Fig A8.2). The absorbed energy causes a rise in temperature of these areas which may be harmful. When damage is caused the result is called chorio-retinal injury[4]. The main effect of a chorio-retinal injury is to destroy the part of the retina under the retinal image of the source of the radiation. The probability of chorio-retinal injury occurring is mainly determined by the retinal radiant exposure, i.e. the product of the retinal irradiance and the exposure time. The retinal irradiance is determined by the nature of the radiant source, the pupil size and the transmission of the ocular media. As for exposure time, this can be conveniently divided into two parts, exposure times less than and greater than 0.25 seconds. The reason for this division is that the normal response to a very bright light is to blink and look away. Such an action takes at least 0.25 seconds but thereafter provides protection. Fortunately very high retinal irradiances are required to produce the damaging radiant exposure within 0.25 seconds.

The mechanism which creates chorio-retinal injury is thermal. In addition to this effect there is also evidence[4] that absorption of visible radiation, particularly of that at the blue end of the visible spectrum, can cause damage to the retinal receptors. This photo-chemical effect also requires high radiant exposures.

#### A8.2.3   Infra-red radiation

The Commission Internationale de l'Eclairage has divided the infra-red region of the electro-magnetic spectrum into three parts, IR-A (780-1400 nm), IR-B (1.4-3.0 $\mu$m) and IR-C (3.0 $\mu$m-1 mm)[1]. The effect of infra-red radiation depends on where the radiation is absorbed. Most of the IR-A radiation is absorbed in the ocular media of the eye but some of it reaches the retina

*Appendix 8 Physiological effects of optical radiation*

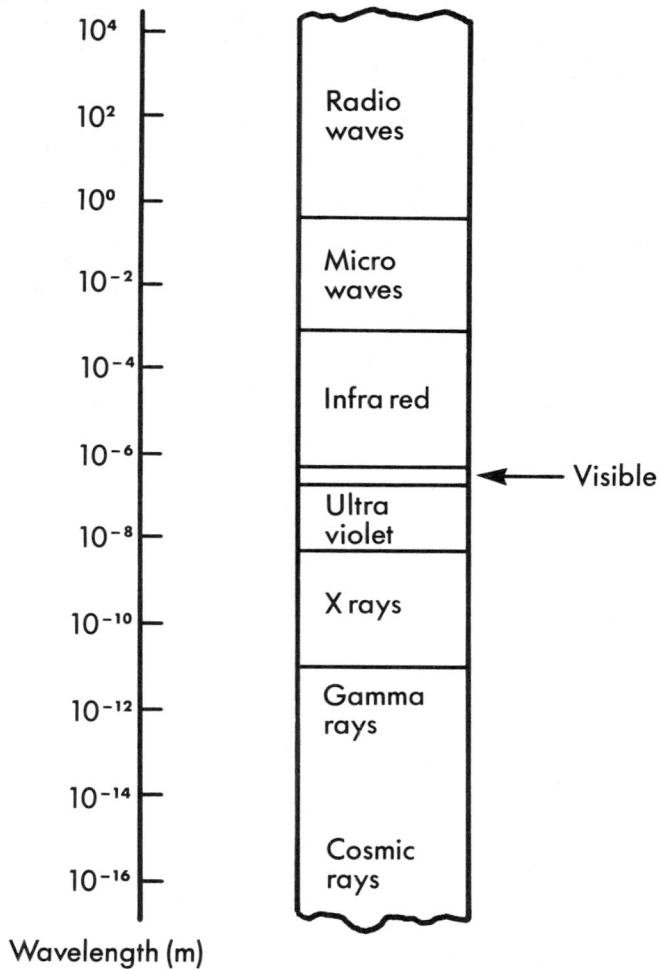

Wavelength (m)

Fig. A8.1 A schematic diagram of the electro-magnetic spectrum

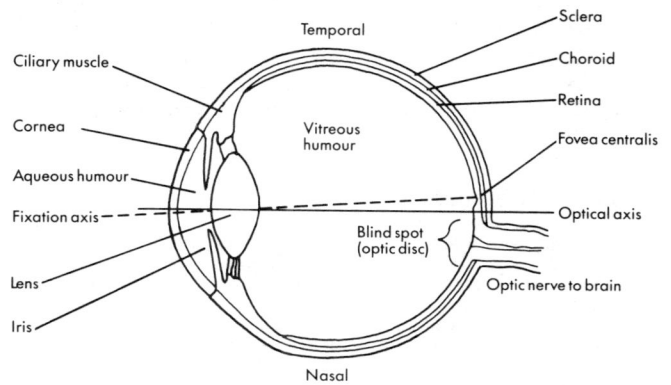

Fig. A8.2 A horizontal section through the human eye

and is absorbed there. IR-B radiation in the wavelength range 1400-1900 nm is almost all absorbed in the cornea and aqueous humour of the eye (Fig A8.2). Infra-red radiation above 1900 nm does not penetrate the eye, being absorbed by the cornea. The effect of the IR-A radiation that reaches the retina has already been discussed as chorio-retinal injury. The IR-A and IR-B radiation that is absorbed in the ocular media or the cornea will raise the temperature of these tissues. It is believed that elevated temperatures in the lens are related to the development of opacities there, particularly if the exposure is prolonged.

As for damage to the skin caused by visible and infra-red radiation, if the temperature elevation caused by the absorption of radiation is sufficient, burns will occur. The irradiance necessary to cause skin burns depends on the area and reflectance of the exposed skin and the duration of the exposure.

## A8.3 Practical aspects

### A8.3.1 Threshold limit values

The above discussion of the effects of optical radiation has shown that it is necessary to protect people from excessive exposure to such radiation. But how much is excessive? The most widely accepted recommendations

about safe exposures to the various types of optical radiation are the threshold limit values published by the American Conference of Governmental Industrial Hygienists (ACGIH)[6]. These threshold limit values quantify, in appropriate forms, the optical radiation conditions to which people may be repeatedly exposed without adverse effect. The relevant threshold limit values are listed in table A8.1. The various weighting functions which are used to specify the spectral sensitivity of the eye or skin to the different hazards are given in table A8.2. It should be noted that the threshold limit values for ultra-violet radiation are established recommendations and are supported by the U.K. National Radiological Protection Board. However, the threshold limit values for visible and near infra-red radiation are provisional and have not yet been adopted.

### A8.3.2 Light sources

By using the recommended threshold limit values listed in table A8.1, it is possible to assess the hazards that various light sources pose to people. The light sources commonly used in interiors, when used in conventional ways, are unlikely to exceed any of these limitations[5,7]. However, there exist several very powerful light sources used for floodlighting as well as several different types of source which are designed as sources of optical radiation rather than light; for example, ultra-violet lamps for the curing of ink, ultra-violet lamps for sunbeds, lamps for infra-red drying. In addition, there occur some applications such as ophthamological examination, in which large amounts of light are projected into the eye[4]. Whenever such lamps or applications are being considered, care should be taken to check that the limits in table A8.1 are not exceeded for realistic conditions of exposure. It should also be noted that the recommendations given in table A8.1 do not apply to lasers. For these sources of optical radiation a different set of threshold limit values are available[6,8].

### A8.3.3 Exposure control

If the limits in table A8.1 are exceeded then the user is left with three choices. The source of the offending radiation can be changed, the method of operation can be changed, or some degree of personal protection can be provided. The first option is the best but it is not always available. The second option is acceptable provided the revised mode of operation is adhered to. Personal protection can take two forms: screening of the source with materials opaque to the hazardous radiation and/or some form of eye filters, helmets, clothing, etc. Advice is available on suitable approaches of this type[9,10].

## A8.4   References

1   Commission Internationale de l'Eclairage, International Lighting Vocabulary, CIE Publication 17, CIE Paris, 1970.
2.  Steck, B. Effects of optical radiation on man, Ltg. Res. & Technol., **14**, 130, 1982.
3   Parrish, J.A., Anderson, R.R., Urbach, F. and Pitts, D. UV-A, Biological effects of ultra-violet radiation with emphasis on human responses to longwave ultra-violet, J. Wiley & Sons, Chichester, 1978.
4   Wolbarsht, M.L. and Sliney, D.H. Ocular effects of non-ionising radiation. Proc. of the Society of Photo-Optical Instrumentation Engineers, Vol. 229, The Society of Photo-Optical Instrumentation Engineers, Washington, U.S.A. 1980.
5   Sliney, D.H. Optical radiation safety, Ltg. Res. & Technol., **14**, 142, 1982.
6   American Conference of Governmental Industrial Hygienists, Threshold limit values for chemical substances and physical agents in the workroom environment with intended changes for 1981, ACGIH, Cincinatti, USA, 1981.
7   Spears, G.R., Radiant flux measurements of ultra-violet emitting light sources. J.I.E.S., **4**, 36, 1974.
8   American National Standards Institute, American National Standard for the safe use of lasers Z136.1, ANSI, New York, 1976.
9   National Radiological Protection Board, Protection against ultra-violet radiation in the workplace, HMSO, London, 1977.
10  Sliney, D.H. Non-ionising radiation, in Industrial Environmental Health, ed. by L.V. Cralley, Academic Press, London, 1972.

**Table A8.1**   ACGIH Threshold Limit Values for ultra-violet, visible and infra-red radiation[5,6]

| Wavelength range | Threshold Limit Value | Notes |
|---|---|---|
| Ultra-violet 320-400 nm | irradiance on skin or eye $\leqslant 10$ W/m$^2$ | For periods >1000 seconds |
| Ultra-violet 320-400 nm | radiant exposure on skin or eye $\leqslant 10,000$ J m$^{-2}$ | For periods <1000 seconds |
| Ultra-violet 200-315 nm | exposure time on skin or eye(s) $\sum\limits_{200}^{315} \dfrac{30}{E_\lambda\ S_\lambda\ \triangle\lambda}$ | where $E_\lambda$ = spectral irradiance at eye of skin (Wm$^{-2}$ nm$^{-1}$) $S_\lambda$ = relative spectral effectiveness (see table A8.2) $\triangle\lambda$ = bandwidth (nm) |
| Visible and near infra-red 400-1400 nm | $\sum\limits_{400}^{1400} L_\lambda\ R_\lambda\ \triangle\lambda \leqslant \alpha^{-1} t^{-0.5} \times 10^4$ | where $L_\lambda$ = spectral radiance of the source in bandwidth $\triangle\lambda$ (W.m$^{-2}$. sr$^{-1}$) $R_\lambda$ = burn hazard fuction (see table A8.2) $\alpha$ = angular subtense of the largest dimension of the source (rad) $t$ = viewing duration (s) (values of $t$ are limited to the range 1$\mu$s to 10s) |
| Visible and near infra-red 400-1400 nm | $\sum\limits_{400}^{1400} L_\lambda\ t B_\lambda\ \triangle\lambda \leqslant 10^6$ J m$^{-2}$sr$^{-1}$ | For $t \leqslant 10,000$ s, sources subtending more than 0.011 rad where $L_\lambda$ = spectral radiance of the source in bandwidth $\triangle\lambda$ (W. m$^{-2}$ sr$^{-1}$) $B_\lambda$ = blue light hazard function (see table A8.2) $t$ = exposure time (s) |
| Visible and near infra-red 400-1400 nm | $\sum\limits_{400}^{1400} E_\lambda\ t B_\lambda\ \triangle\lambda \leqslant 100$ J/m$^2$ | For $t \leqslant 10,000$ s, sources subtending less than 0.011 rads where $E_\lambda$ = spectral irradiance at the eye (W m$^{-2}$ nm$^{-1}$) $B_\lambda$ = blue light hazard function (see table A8.2) $t$ = exposure time (s) $\triangle\lambda$ = bandwidth (nm) |
| Visible and near infra-red 400-1400 nm | $\sum\limits_{400}^{1400} L_\lambda\ B_\lambda\ \triangle\lambda \leqslant 100$ W m$^{-2}$sr$^{-1}$ | For $t > 10,000$ s, and sources subtending more than 0.011 rad where $L_\lambda$ = spectral radiance of the source in bandwidth $\triangle\lambda$(W m$^{-2}$ sr$^{-1}$) $B_\lambda$ = blue light hazard function (see table A8.2) |
| Visible and near infra-red 400-1400 nm | $\sum\limits_{400}^{1400} E_\lambda\ B_\lambda\ \triangle\lambda \leqslant 0.01$ W/m$^2$ | For $t \leqslant 10,000$ s, and sources subtending less than 0.011 rad where $E_\lambda$ = spectral irradiance at the eye (W m$^{-2}$ nm$^{-1}$) $B_\lambda$ = blue light hazard function (see table A8.2) $\triangle\lambda$ = bandwidth (nm) |
| Near infra-red >770 nm | Irradiance at eye $\leqslant 100$ W/m$^2$ | For exposures longer than 100 s |
| Near infra-red 770-1400 nm | $\sum\limits_{770}^{1400} L_\lambda\ \triangle\lambda = \dfrac{0.6 \times 10^4}{\alpha}$ | For infra-red lamps where strong visual stimulus is absent, and exposure is longer than 100 s. where $L_\lambda$ = spectral radiance of the source in bandwidth $\triangle\lambda$ (W m$^{-2}$ sr$^{-1}$) $\alpha$ = angular subtense of source (rad) |

**Table A8.2**   Weighting functions for ACGIH recommendations in table A8.1[6]

| Wavelength (nm) | $S_\lambda$ = Relative spectral effectiveness | Wavelength (nm) | $R_\lambda$ = Burn hazard function | $B_\lambda$ = Blue light hazard function |
|---|---|---|---|---|
| 200 | 0.03 | 400 | 1.0 | 0.10 |
| 210 | 0.075 | 405 | 2.0 | 0.20 |
| 220 | 0.12 | 410 | 4.0 | 0.40 |
| 230 | 0.19 | 415 | 8.0 | 0.80 |
| 240 | 0.30 | 420 | 9.0 | 0.90 |
| 250 | 0.43 | 425 | 9.5 | 0.95 |
| 254 | 0.50 | 430 | 9.8 | 0.98 |
| 260 | 0.65 | 435 | 10.0 | 1.00 |
| 270 | 1.00 | 440 | 10.0 | 1.00 |
| 280 | 0.88 | 445 | 9.7 | 0.97 |
| 290 | 0.64 | 450 | 9.4 | 0.94 |
| 300 | 0.30 | 455 | 9.0 | 0.90 |
| 305 | 0.06 | 460 | 8.0 | 0.80 |
| 310 | 0.015 | 465 | 7.0 | 0.70 |
| 315 | 0.003 | 470 | 6.2 | 0.62 |
| – | – | 475 | 5.5 | 0.55 |
| – | – | 480 | 4.5 | 0.45 |
| – | – | 485 | 4.0 | 0.40 |
| – | – | 490 | 2.2 | 0.22 |
| – | – | 495 | 1.6 | 0.16 |
| – | – | 500–600 | 1.0 | $10^{(450-\lambda)/50}$ |
| – | – | 600–700 | 1.0 | 0.001 |
| – | – | 700–1 060 | $10.0^{(700-\lambda)/505}$ | 0.001 |
| – | – | 1 060–1 400 | 0.2 | 0.001 |

## APPENDIX 9   PREDICTING AND MEASURING DAYLIGHT FACTOR

### A9.1   Introduction

The daylight factor at a point indoors is defined as the illuminance received at that point from a sky of known or assumed luminance distribution expressed as a percentage of the horizontal illuminance outdoors from an unobstructed hemisphere of the same sky. Direct sunlight is excluded from both illuminances. Usually the sky assumed is the overcast sky specified by the CIE[1].

The daylight factor at a point in an interior may be considered to consist of a direct component entering through the window and an indirect component received at the point after reflection from the internal surfaces.

The direct component may conveniently be treated as the sum of two other components, (a) the sky component which arrives directly from the sky, (b) the externally reflected component received directly after reflection from outdoor surfaces, such as the wall of a building opposite. Thus the total daylight factor at a point is given by the sum of the sky component, the externally reflected component and the internally reflected component, each component being expressed as a percentage of the outdoor horizontal illuminance from an unobstructed hemisphere of a sky of known or assumed luminance distribution.

**Table A9.1**   Sky Components (CIE Standard Overcast Sky) for Vertical Glazed Rectangular Windows

Ratio $H/D$ = height of window head above working plane: distance from window

| $H/D$ | 0.1 | 0.2 | 0.3 | 0.4 | 0.5 | 0.6 | 0.7 | 0.8 | 0.9 | 1.0 | 1.2 | 1.4 | 1.6 | 1.8 | 2.0 | 2.5 | 3.0 | 4.0 | 6.0 | ∞ | Angle of obstruction |
|---|---|---|---|---|---|---|---|---|---|---|---|---|---|---|---|---|---|---|---|---|---|
| ∞ | 1.3 | 2.5 | 3.7 | 4.9 | 5.9 | 6.9 | 7.7 | 8.4 | 9.0 | 9.6 | 10.7 | 11.6 | 12.2 | 12.6 | 13.0 | 13.7 | 14.2 | 14.6 | 14.9 | 15.0 | 90° |
| 5.0 | 1.2 | 2.4 | 3.7 | 4.8 | 5.9 | 6.8 | 7.6 | 8.3 | 8.8 | 9.4 | 10.5 | 11.1 | 11.7 | 12.3 | 12.7 | 13.3 | 13.7 | 14.0 | 14.1 | 14.2 | 79° |
| 4.0 | 1.2 | 2.4 | 3.6 | 4.7 | 5.8 | 6.7 | 7.4 | 8.2 | 8.7 | 9.2 | 10.3 | 10.9 | 11.4 | 12.0 | 12.4 | 12.9 | 13.3 | 13.5 | 13.6 | 13.7 | 76° |
| 3.5 | 1.2 | 2.4 | 3.6 | 4.6 | 5.7 | 6.6 | 7.3 | 8.0 | 8.5 | 9.0 | 10.1 | 10.6 | 11.1 | 11.8 | 12.2 | 12.6 | 12.9 | 13.2 | 13.2 | 13.3 | 74° |
| 3.0 | 1.2 | 2.3 | 3.5 | 4.5 | 5.5 | 6.4 | 7.1 | 7.8 | 8.2 | 8.7 | 9.8 | 10.2 | 10.7 | 11.3 | 11.7 | 12.0 | 12.4 | 12.5 | 12.6 | 12.7 | 72° |
| 2.8 | 1.1 | 2.3 | 3.4 | 4.5 | 5.4 | 6.3 | 7.0 | 7.6 | 8.1 | 8.6 | 9.6 | 10.0 | 10.5 | 11.1 | 11.4 | 11.7 | 12.0 | 12.2 | 12.3 | 12.3 | 70° |
| 2.6 | 1.1 | 2.2 | 3.4 | 4.4 | 5.3 | 6.2 | 6.8 | 7.5 | 7.9 | 8.4 | 9.3 | 9.8 | 10.2 | 10.8 | 11.1 | 11.4 | 11.7 | 11.8 | 11.9 | 11.9 | 69° |
| 2.4 | 1.1 | 2.2 | 3.3 | 4.3 | 5.2 | 6.0 | 6.6 | 7.3 | 7.7 | 8.1 | 9.1 | 9.5 | 10.0 | 10.4 | 10.7 | 11.0 | 11.2 | 11.3 | 11.4 | 11.5 | 67° |
| 2.2 | 1.1 | 2.1 | 3.2 | 4.1 | 5.0 | 5.8 | 6.4 | 7.0 | 7.4 | 7.9 | 8.7 | 9.1 | 9.6 | 10.0 | 10.2 | 10.5 | 10.7 | 10.8 | 10.9 | 10.9 | 66° |
| 2.0 | 1.0 | 2.0 | 3.1 | 4.0 | 4.8 | 5.6 | 6.2 | 6.7 | 7.1 | 7.5 | 8.3 | 8.7 | 9.1 | 9.5 | 9.7 | 9.9 | 10.0 | 10.1 | 10.2 | 10.3 | 63° |
| 1.9 | 1.0 | 2.0 | 3.0 | 3.9 | 4.7 | 5.4 | 6.0 | 6.5 | 6.9 | 7.3 | 8.1 | 8.5 | 8.8 | 9.2 | 9.4 | 9.6 | 9.7 | 9.8 | 9.9 | 9.9 | 62° |
| 1.8 | 0.97 | 1.9 | 2.9 | 3.8 | 4.6 | 5.3 | 5.8 | 6.3 | 6.7 | 7.1 | 7.8 | 8.2 | 8.5 | 8.8 | 9.0 | 9.2 | 9.3 | 9.4 | 9.5 | 9.5 | 61° |
| 1.7 | 0.94 | 1.9 | 2.8 | 3.6 | 4.4 | 5.1 | 5.6 | 6.1 | 6.5 | 6.8 | 7.5 | 7.8 | 8.2 | 8.5 | 8.6 | 8.8 | 8.9 | 9.0 | 9.1 | 9.1 | 60° |
| 1.6 | 0.90 | 1.8 | 2.7 | 3.5 | 4.2 | 4.9 | 5.4 | 5.8 | 6.2 | 6.5 | 7.2 | 7.5 | 7.8 | 8.1 | 8.2 | 8.4 | 8.5 | 8.6 | 8.6 | 8.6 | 58° |
| 1.5 | 0.86 | 1.7 | 2.6 | 3.3 | 4.0 | 4.6 | 5.1 | 5.6 | 5.9 | 6.2 | 6.8 | 7.1 | 7.4 | 7.6 | 7.8 | 7.9 | 8.0 | 8.0 | 8.1 | 8.1 | 56° |
| 1.4 | 0.82 | 1.6 | 2.4 | 3.2 | 3.8 | 4.4 | 4.8 | 5.2 | 5.6 | 5.9 | 6.4 | 6.7 | 7.0 | 7.2 | 7.3 | 7.4 | 7.5 | 7.5 | 7.6 | 7.6 | 54° |
| 1.3 | 0.77 | 1.5 | 2.3 | 2.9 | 3.6 | 4.1 | 4.5 | 4.9 | 5.2 | 5.5 | 5.9 | 6.2 | 6.4 | 6.6 | 6.7 | 6.8 | 6.9 | 6.9 | 6.9 | 7.0 | 52° |
| 1.2 | 0.71 | 1.4 | 2.1 | 2.7 | 3.3 | 3.8 | 4.2 | 4.5 | 4.8 | 5.0 | 5.4 | 5.7 | 5.9 | 6.0 | 6.1 | 6.2 | 6.2 | 6.3 | 6.3 | 6.3 | 50° |
| 1.1 | 0.65 | 1.3 | 1.9 | 2.5 | 3.0 | 3.4 | 3.8 | 4.1 | 4.3 | 4.6 | 4.9 | 5.1 | 5.3 | 5.4 | 5.4 | 5.5 | 5.6 | 5.6 | 5.7 | 5.7 | 48° |
| 1.0 | 0.57 | 1.1 | 1.7 | 2.2 | 2.6 | 3.0 | 3.3 | 3.6 | 3.8 | 4.0 | 4.3 | 4.5 | 4.6 | 4.7 | 4.7 | 4.8 | 4.8 | 4.9 | 5.0 | 5.0 | 45° |
| 0.9 | 0.50 | 0.99 | 1.5 | 1.9 | 2.2 | 2.6 | 2.8 | 3.1 | 3.3 | 3.4 | 3.7 | 3.8 | 3.9 | 4.0 | 4.0 | 4.0 | 4.1 | 4.1 | 4.2 | 4.2 | 42° |
| 0.8 | 0.42 | 0.83 | 1.2 | 1.6 | 1.9 | 2.2 | 2.4 | 2.6 | 2.7 | 2.9 | 3.1 | 3.2 | 3.3 | 3.3 | 3.3 | 3.3 | 3.4 | 3.4 | 3.4 | 3.1 | 39° |
| 0.7 | 0.33 | 0.68 | 0.97 | 1.3 | 1.5 | 1.7 | 1.9 | 2.1 | 2.2 | 2.3 | 2.5 | 2.5 | 2.6 | 2.6 | 2.6 | 2.6 | 2.7 | 2.7 | 2.8 | 2.8 | 35° |
| 0.6 | 0.24 | 0.53 | 0.74 | 0.98 | 1.2 | 1.3 | 1.5 | 1.6 | 1.7 | 1.8 | 1.9 | 1.9 | 2.0 | 2.0 | 2.0 | 2.1 | 2.1 | 2.1 | 2.1 | 2.1 | 31° |
| 0.5 | 0.16 | 0.39 | 0.52 | 0.70 | 0.82 | 0.97 | 1.0 | 1.10 | 1.2 | 1.3 | 1.4 | 1.4 | 1.4 | 1.4 | 1.5 | 1.5 | 1.5 | 1.5 | 1.5 | 1.5 | 27° |
| 0.4 | 0.10 | 0.25 | 0.34 | 0.45 | 0.54 | 0.62 | 0.70 | 0.75 | 0.82 | 0.89 | 0.92 | 0.95 | 0.95 | 0.96 | 0.96 | 0.96 | 0.97 | 0.97 | 0.98 | 0.98 | 22° |
| 0.3 | 0.06 | 0.14 | 0.18 | 0.26 | 0.30 | 0.34 | 0.38 | 0.42 | 0.44 | 0.47 | 0.49 | 0.50 | 0.50 | 0.51 | 0.51 | 0.52 | 0.52 | 0.52 | 0.53 | 0.53 | 17° |
| 0.2 | 0.03 | 0.06 | 0.09 | 0.11 | 0.12 | 0.14 | 0.16 | 0.20 | 0.21 | 0.21 | 0.22 | 0.22 | 0.22 | 0.22 | 0.23 | 0.23 | 0.23 | 0.23 | 0.24 | 0.24 | 11° |
| 0.1 | 0.01 | 0.02 | 0.02 | 0.03 | 0.03 | 0.04 | 0.04 | 0.05 | 0.05 | 0.05 | 0.06 | 0.06 | 0.06 | 0.06 | 0.07 | 0.07 | 0.07 | 0.07 | 0.08 | 0.08 | 6° |
| 0 | 0.1 | 0.2 | 0.3 | 0.4 | 0.5 | 0.6 | 0.7 | 0.8 | 0.9 | 1.0 | 1.2 | 1.4 | 1.6 | 1.8 | 2.0 | 2.5 | 3.0 | 4.0 | 6.0 | ∞ | 0° |

Ratio $W/D$ = width of window to one side of normal: distance from window

## Table A9.2 Minimum Internally Reflected Component of Daylight Factor

| Ratio of Window Area to Floor Area | Window Area as Percentage of Floor Area | Floor Reflectance (%) 10 | | | | Floor Reflectance (%) 20 | | | | Floor Reflectance (%) 40 | | | |
|---|---|---|---|---|---|---|---|---|---|---|---|---|---|
| | | Wall Reflectance (%) | | | | Wall Reflectance (%) | | | | Wall Reflectance (%) | | | |
| | | 20 | 40 | 60 | 80 | 20 | 40 | 60 | 80 | 20 | 40 | 60 | 80 |
| | | (%) | (%) | (%) | (%) | (%) | (%) | (%) | (%) | (%) | (%) | (%) | (%) |
| 1:50 | 2 | – | – | 0.1 | 0.2 | – | 0.1 | 0.1 | 0.2 | – | 0.1 | 0.2 | 0.2 |
| 1:20 | 5 | 0.1 | 0.1 | 0.2 | 0.4 | 0.1 | 0.2 | 0.3 | 0.5 | 0.1 | 0.2 | 0.4 | 0.6 |
| 1:14 | 7 | 0.1 | 0.2 | 0.3 | 0.5 | 0.1 | 0.2 | 0.4 | 0.6 | 0.2 | 0.3 | 0.6 | 0.8 |
| 1:10 | 10 | 0.1 | 0.2 | 0.4 | 0.7 | 0.2 | 0.3 | 0.6 | 0.9 | 0.3 | 0.5 | 0.8 | 1.2 |
| 1:6.7 | 15 | 0.2 | 0.4 | 0.6 | 1.0 | 0.2 | 0.5 | 0.8 | 1.3 | 0.4 | 0.7 | 1.1 | 1.7 |
| 1:5 | 20 | 0.2 | 0.5 | 0.8 | 1.4 | 0.3 | 0.6 | 1.1 | 1.7 | 0.5 | 0.9 | 1.5 | 2.3 |
| 1:4 | 25 | 0.3 | 0.6 | 1.0 | 1.7 | 0.4 | 0.8 | 1.3 | 2.0 | 0.6 | 1.1 | 1.8 | 2.8 |
| 1:3.3 | 30 | 0.3 | 0.7 | 1.2 | 2.0 | 0.5 | 0.9 | 1.5 | 2.4 | 0.8 | 1.3 | 2.1 | 3.3 |
| 1:2.9 | 35 | 0.4 | 0.8 | 1.4 | 2.3 | 0.5 | 1.0 | 1.8 | 2.8 | 0.9 | 1.5 | 2.4 | 3.8 |
| 1:2.5 | 40 | 0.5 | 0.9 | 1.6 | 2.6 | 0.6 | 1.2 | 2.0 | 3.1 | 1.0 | 1.7 | 2.7 | 4.2 |
| 1:2.2 | 45 | 0.5 | 1.0 | 1.8 | 2.9 | 0.7 | 1.3 | 2.2 | 3.4 | 1.2 | 1.9 | 3.0 | 4.6 |
| 1:2 | 50 | 0.6 | 1.1 | 1.9 | 3.1 | 0.8 | 1.4 | 2.3 | 3.7 | 1.3 | 2.1 | 3.2 | 4.9 |

Note: Ceiling reflectance assumed = 70%, floor area assumed = 36 m², ceiling height assumed = 3m.
A long external obstruction of 20° measured from the mid-height of the window is assumed as a common condition.

## A9.2 Calculation of daylight factor for windows

Methods of point by point daylight factor calculation for windows are legion[2,3]. Only two simple methods will be considered here. More elaborate methods such as the Waldram diagram[4] and the BRE protractors[4,5] are available when greater accuracy is required or where complex window shapes or non-vertical glazing are being considered.

### A9.2.1 BRE tabular method

Perhaps the simplest way of determining the daylight factor at a point indoors for windows with clear, vertical, rectangular glazing in conjunction with a CIE standard overcast sky[1], is to use Tables A9.1 to A9.5[4]. Table A9.1 will provide the sky component . To use the table the following information is required:

$H_1$, $H_2$ – the heights of the window head and sill above the working plane.
$W_1$, $W_2$ – the distances of the window's vertical edges from a line drawn from the reference point for which the daylight factor is to be calculated, normal to the plane of the window.
$D$ – the distance from the reference point to the plane of the window (This is the plane of the inside of the wall or the outside whichever edge of the window aperture limits the view of the sky.)

Given the appropriate values for each of these distances, the sky component can be obtained from Table A9.1. For example, Fig A9.1 shows the procedure for a window with a sill on the working plane and with the reference point and on the centre line of the window. For this situation let $H_1 = 8$, $H_2 = 0$, $W_1 = 12$, $W_2 = 12$, $D = 10$. Therefore $H_1/D = 0.8$, $H_2/D = 0$, $W_1/D = 1.2$, $W_2/D = 1.2$. From Table 9.1 the sky component of window area $P = 3.1\%$, the sky component of window area $Q = 3.1\%$, therefore the sky component of window $PQ = 6.2\%$.

In general, the sky component at any reference point can be obtained by addition or subtraction of sky components for appropriate areas. For example, Fig A9.2 shows this procedure for a window with the sill above the working plane and with the reference point off

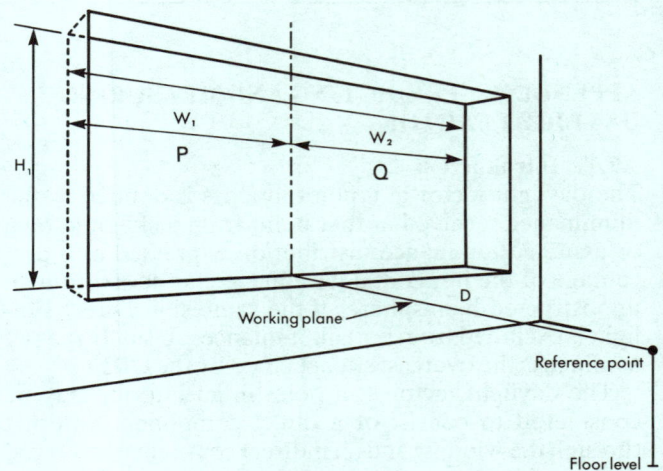

**Fig. A9.1** A window with the sill on the working plane, and the reference point on the centre line of the window

centre, outside the width of the window. For this situation let $H_1 = 8$, $H_2 = 2$, $W_1 = 8$, $W_2 = 24$, $D = 10$. Therefore $H_1/D = 0.8$, $H_2/D = 0.2$, $W_1/D = 0.8$, $W_2/D = 2.4$. Then from Table A9.1 the sky component of window area $PQRS = 3.3\%$, the sky component of window area $PR = 2.6\%$, the sky component of window area $RS = 0.23\%$, the sky component of window area $R = 0.20\%$. Therefore the sky component of window $Q$ equals the sky component of $PQRS$ minus the sky component of $PR$ minus sky component of $RS$ plus sky component of $R$,

$Q = PQRS - PR - RS + R$
Therefore,
$Q = 3.3 - 2.6 - 0.23 + 0.20 = 0.67\%$.

If the direct entry of light through the window is severely limited by an external obstruction, it will be necessary to calculate the externally reflected component $ERC$. This can also be done using Table A9.1. The procedure is to treat the external obstruction visible from the reference point as a patch of sky whose luminance is some fraction of that of the sky obscured. In other words, the sky component for the obstructed area is first calculated as described above and is then converted to the externally reflected component by multiplying by the ratio of the

*Appendix 9 Predicting and measuring daylight factor*

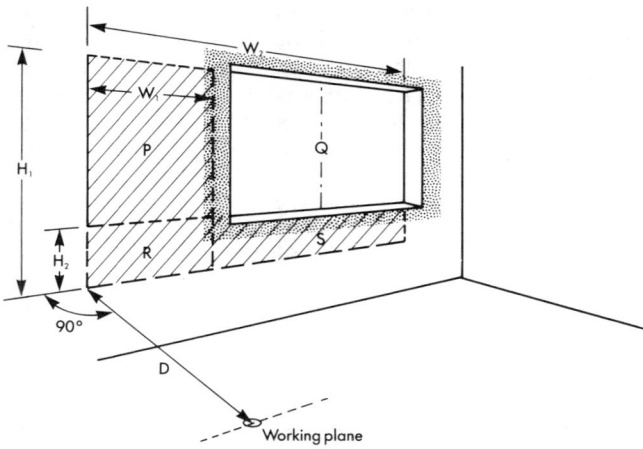

**Fig. A9.2** A window with a high sill, and the reference point outside the width of the window

luminance of the obstructed area to the sky luminance.

Of course, if the external obstruction is sufficient to justify calculating the externally reflected component, it will also be necessary to modify the window area for which the sky component is calculated. Specifically, the average height of the external obstruction viewed from the reference point should be treated as the effective window sill level. Fig A9.3 demonstrates the procedure for calculating the sky component and externally reflected component for a window which is subject to external obstruction. In Fig A9.3 the window sill is on the horizontal plane containing the reference point and the external obstruction is bounded by the roof line.

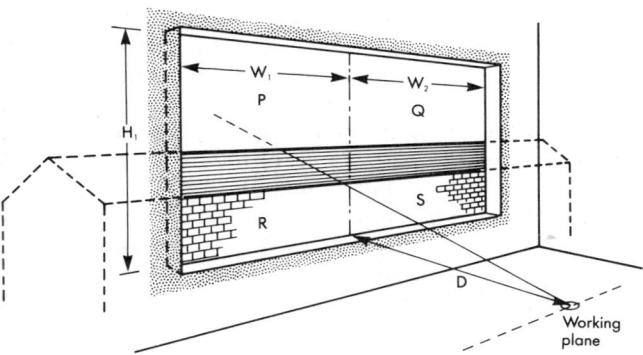

**Fig. A9.3** A window with the sill on the working plane, and the reference point on the centre line of the window. A continuous external obstruction is parallel to the window

The sky component is calculated for window area $PQ$. For window $PQRS$, let $H_1 = 8, H_2 = 0, W_1 = 12, W_2 = 12, D = 10$. Therefore $H_1/D = 0.8, H_2/D = 0, W_1/D = 1.2, W_2/D = 1.2$. From Table A9.1 the sky component of window $PR = 3.1\%$. The sky component of window $QS = 3.1\%$. Therefore the sky component of window $PQRS = 3.1 + 3.1 = 6.2\%$. For window $RS$, let $H_1 = 4, H_2 = 0, W_1 = 12, W_2 = 12, D = 10$. Therefore $H_1/D = 0.4, H_2/D = 0, W_1/D = 1.2, W_2/D = 1.2$. From Table A9.1, the sky component of window $RS = 0.92 + 0.92 = 1.84\%$. Therefore the sky component of window $PQ = $ the sky component of window $PQRS$ – the sky component of window $RS = 6.20 - 1.84 = 4.36\%$.

The externally reflected component is calculated for window $RS$. Assuming the luminance of the obstruction is one-tenth of the luminance of the sky it obstructs, the externally reflected component for window $RS = $

$$RS \times \frac{\text{luminance of obstruction}}{\text{luminance of sky}} = 1.84 \times 0.1 = 0.184\%$$

Therefore for window area $PQRS$, the sky component = $4.36\%$, externally reflected component = $0.18\%$.

The minimum internally reflected component (IRC) can be read from Table A9.2 for a typical, medium sized, square room with a single window in one wall. It can be converted to other room size and ceiling reflectances by multiplying the IRC from Table 9.2 by the appropriate factors from Tables A9.3 and A9.4. By using Table A9.5 conversion may be made to the average IRC throughout the interior, by multiplying the minimum IRC by the appropriate factor. (An alternative and more precise method of allowing for the internally reflected component in the calculation of daylight factor is given at the end of section A9.2.2).

Having calculated the sky component, the externally reflected component (where applicable) and the internally reflected component for a reference point, the day light factor can be obtained by simple summation. If more than one window is involved, the daylight factor should be calculated for each window and the total daylight factor obtained by summation.

**Table A9.3** Conversion Factors for Rooms where Floor Area Corresponds to 10 m² and 100 m²

| Floor Area (m²) | Wall Reflectance | | | |
|---|---|---|---|---|
| | 20% | 40% | 60% | 80% |
| 10 | 0.6 | 0.7 | 0.8 | 0.9 |
| 100 | 1.4 | 1.2 | 1.0 | 0.9 |

**Table A9.4** Conversion Factors for Ceiling Reflectance

| Ceiling Reflectance (%) | Conversion Factor |
|---|---|
| 40 | 0.7 |
| 50 | 0.8 |
| 60 | 0.9 |
| 70 | 1.0 |
| 80 | 1.1 |

**Table A9.5** Conversion Factors for Average Internally Reflected Component

| Wall Reflectance (%) | Conversion Factor |
|---|---|
| 20 | 1.8 |
| 40 | 1.4 |
| 60 | 1.3 |
| 80 | 1.2 |

It should be noted that the values given in Table A9.1 assume a window with clean, clear, glass. Therefore allowance should be made in the daylight factor for (a) the effects of glass transmission where other types of glass are used, (b) dirt on the glass and (c) obstruction by window frames and glazing bars. Details of appropriate correction factors are given in reference 4.

Once the daylight factor has been calculated the illuminance at the reference point can be estimated by assuming a value for the horizontal illuminance from an unobstructed sky. Field data[6] reveals that the illuminance from an unobstructed sky in the U.K. can reach 50,000 lux (excluding sunlight). However the value conventionally used for design is 5,000 lux which is exceeded for about 85% of the working year.

### A9.2.2 Perspective drawings in conjunction with a 'pepperpot' diagram

An alternative method of predicting the sky component and externally reflected component for clear, single, vertical glazing in conjunction with a CIE standard overcast sky is to use a perspective drawing and the 'pepperpot' diagram[2]. Figures A9.4a and b show the pepperpot diagram and its use. In order to use the pepperpot diagram, it is necessary to have a perspective of the window and the view through it from the point for

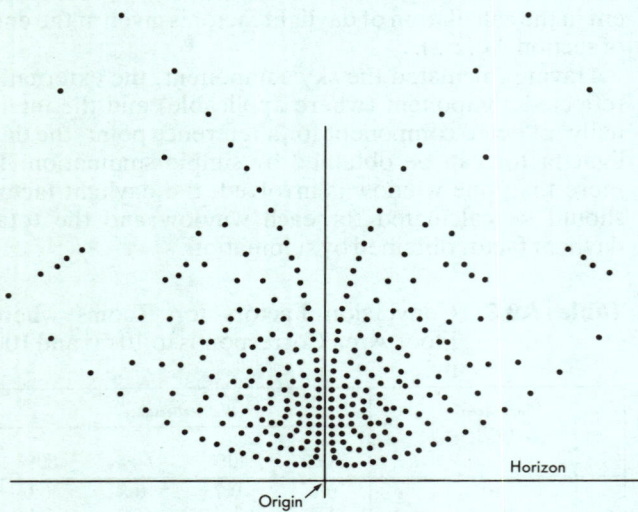

**Fig. A9.4 (a)** A pepperpot diagram for calculating the sky component on a horizontal plane; for use with window perspectives drawn on a vertical picture plane parallel to the window. Each dot denotes a sky component of 0.1%

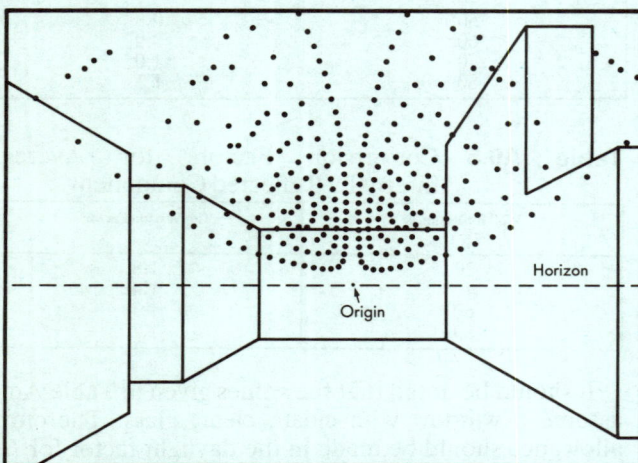

**Fig. A9.4 (b)** An example showing the use of the pepperpot diagram to calculate the sky component and the externally reflected component

which the sky component and the externally reflected component are required. Consider Fig A9.5; let $X$, $Y$ and $Z$ be the cartesian coordinates of say a corner of the vertical window. The coordinates $x$ and $y$ of the corresponding point on the perspective, on a vertical picture plane parallel to the window, must be

$$x = \frac{X.D}{Z}$$

$$y = \frac{Y.D}{Z}$$

where $D$ is the perpendicular distance from the viewpoint to the picture plane, known as the perspective distance. By the use of these relationships it is a simple matter to plot on a sheet of graph paper a perspective of a window and of the scene beyond, seen from a chosen point on the working plane. A full treatment of the geometry of perspective as applied to lighting has recently been published[7] but the following notes will suggest sufficient shortcuts for most lighting purposes: (a) so long as the window is parallel to the picture plane, the perspective of the window will have the same shape as the real window; (b) the perspective of a straight line is always a straight line; (c) if two or more lines are parallel, they will converge towards a common vanishing point in the perspective. If they are parallel to the picture plane, this vanishing point will be at infinity, so the lines will be parallel in the perspective too.

In using perspectives for daylight studies, it is convenient to adopt a single perspective distance such as 20mm. Then, to find the sky component of a window, at a chosen point on the working plane, take that point as the viewpoint for a perspective of the window and the view outside. Place a tracing of the window perspective over a suitably scaled pepperpot chart (Fig A9.4b) making sure that the origin of the pepperpot diagram coincides with the origin ($x = 0$, $y = 0$) of the perspective. Count how many dots $P$ fall within the outline of the sky visible through the window. Each dot represents a sky component of 0.1%, so the total sky component is equal to $P/10$ per cent.

To find the externally reflected component, count how many dots $Q$, fall within the outline of exterior obstructions visible through the window. Each of these dots now represents an externally reflected component of about 0.01%, so the total externally reflected component is roughly equal to $Q/100$ percent. This assumes that the luminance of the obstructions is just one-tenth the luminance of the sky they conceal, which is reasonable for average reflectances of about 0.2 to 0.3. For widely different reflectances, the externally reflected component can be raised or lowered pro rata. For windows in more than one wall, more than one perspective is needed.

To allow for the internally reflected component, information is required on the room proportions, the room surface reflectances and the extent of any external obstruction of the sky. Then, using the method described in reference 2, the total daylight factor is given by the expression

Daylight factor = $a(SC + ERC) + (v. e. g/f)$

where $SC$ = the sky component

$\quad ERC$ = the externally reflected component

$\quad g/f$ = the ratio of the glazing area to the floor area

and $a$, $v$ and $e$ = constants obtained from Tables A9.6, A9.7 and A9.8 respectively.

It should be noted that this formula for daylight factor can also be used with the sky component and externally reflected component calculated by the BRE tabular method.

Again, the values of the sky component and the externally reflected component as found by this method are related to a window with clean, clear, glass. Therefore, allowance should be made in the daylight factor for (a) the effect of glass transmission where other types of glass are used, (b) dirt on the glass and (c) obstruction by window framing and glazing bars. Details of appropriate correction factors are given in reference 4.

### A9.3 Calculation of daylight factor for rooflights

Most rooflight installations are designed to give uniform lighting across a horizontal plane. For such installations there is little point in calculating anything other than the average daylight factor. An expression for the average daylight factor on a horizontal reference plane is given in Section 4.4.1.2 of this Code. More detailed consideration of the daylight factors associated with different forms of rooflights is given in references 2 and 4.

### A9.4 Non-Overcast skies

Daylight factors obtained by the methods discussed above are based on the CIE standard overcast sky[1]. Many other sky compositions occur so a daylight factor based on an overcast sky is of limited usefulness. Nonetheless, it is good practice to design windows to provide the desired visual conditions on an overcast day for, if daylight is adequate on a dull day, it will usually be sufficient when the sun shines. However, the perfectly overcast sky is less suitable for predicting energy use and for optimising lighting control systems; here one should ideally design for an average sky. The average sky at a location represents the mean of a typical succession of real skies, including all the different weather conditions, for each particular sun position. Mathematical models for the average sky have been develped at the Building Research Station[8], the University of Sheffield[9], and the Technical University of Berlin[10].

Under non-overcast sky conditions, window orientation and the varying position of the sun can make a big difference to the amount of daylight received inside a room, and hence to the hours of artificial lighting use. One of the advantages of the average sky is that it can model these effects. Research is continuing into ways in which the average sky can be included in calculations of lighting and energy use prediction[11,12]. It is anticipated that in a few years' time, the average sky approach will be established sufficiently firmly to offer a sound basis for comparing alternative lighting control systems. However, until this goal is achieved, it is recommended that for estimating energy savings, daylight factors be

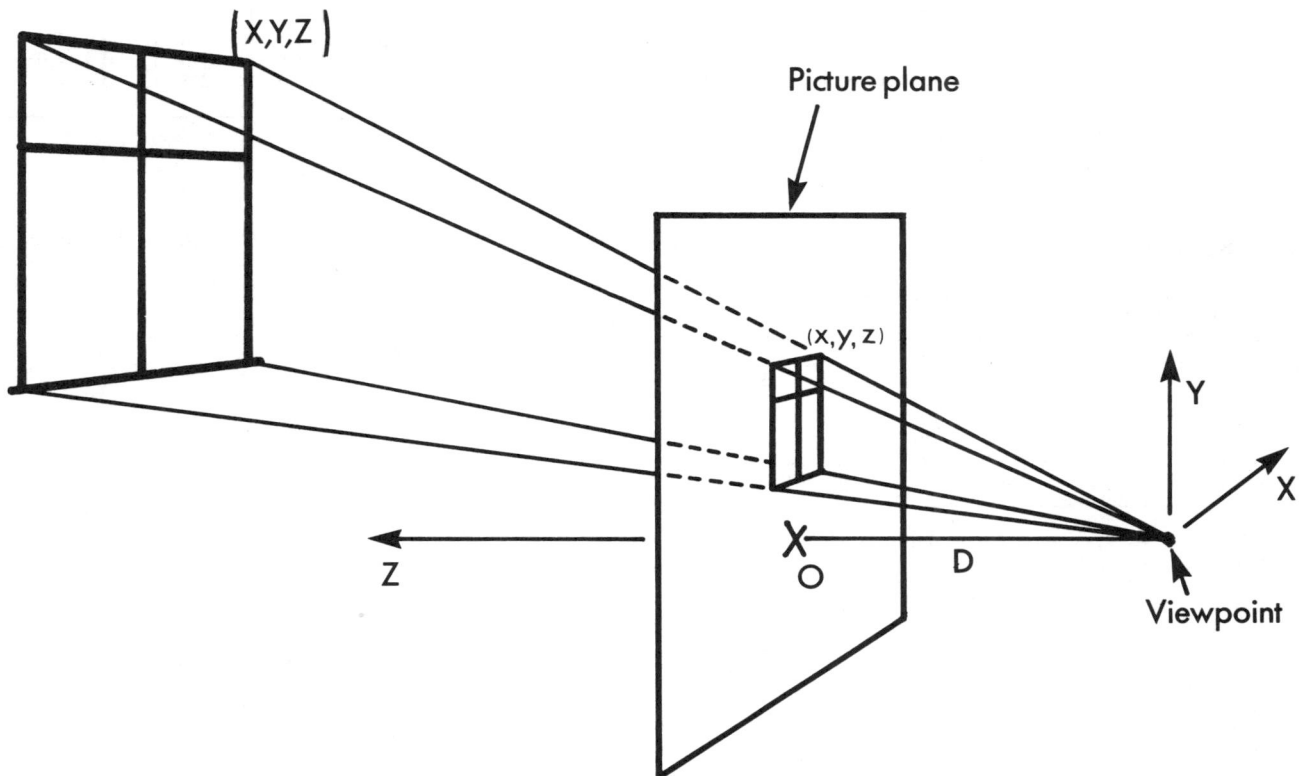

**Fig. A9.5** The relationship between co-ordinates on a perspective and co-ordinates of a real window from the view point

weighted by the orientation factors in Table A9.9, i.e. orientation-weighted daylight factor equal to orientation factor multiplied by overcast sky daylight factor. Intermediate orientation factors should be used for intermediate orientations.

It should be recognised that figures 4.9, 4.10 and 4.11 in Part 4 of this Code have been empirically derived using orientation weighted, overcast sky daylight factors. In general, the use of orientation weighted, overcast sky daylight factors will give pessimistic estimates of energy consumption for automatic photo-electric control systems where side windows are used and optimistic estimates where rooflights are used.

## A9.5   Measurement of daylight factor

If the interior under consideration is already in existence then the daylight factor can be measured using a daylight factor meter[7] at the point under investigation. It will be necessary for the measurements to be made under an overcast sky condition. The practical problems of daylight factor measurement are discussed in reference 2.

## A9.6   Measurement of daylight factor in a model

As an alternative to the calculation of daylight factor, particularly where a complex window design is to be used, a scale model of the interior can be used in conjunction with a real or artificial sky[2,13]. This also has the advantage that it enables a visual assessment of the effect of the daylight to be obtained[2].

The scale model has to be accurately made from opaque materials with the correct internal surface reflectances. Also an accurate representation of any external obstruction must be included. The model is placed under a real or an artifical sky which has a luminance distribution as for the CIE standard overcast sky[1] and measurements of illuminance are made both inside the model, at the point under investigation, and outside, to obtain a measure of the daylight factor, (Fig A9.6). Allowance will need to be made for dirt disposition and for the transmittance of the glazing material unless the model is correctly glazed.

## A9.7   References

[1] Commission Internationale de l'Eclairage, Proc. CIE 13th Session, Zurich, Committee 3.2, CIE Paris, 1955.
[2] Lynes, J.A. Principles of Natural Lighting, Applied Science Publishers, London, 1968.
[3] Hopkinson, R.G., Petherbridge, P. & Longmore, J., Daylighting, Heinemann, London, 1966.
[4] British Standards Institution, BS DD73 1982 – Basic data for the design of buildings: Daylight, BSI, London, 1982.
[5] Longmore, J., BRS Daylight Protractors, HMSO, London, 1968.
[6] Hunt, D.R.G., Improved daylight data for predicting energy savings from photoelectric controls, Ltg. Res and Technol., 11, 9, 1979.
[7] Waldram, J.M. A Manual of Perspective for Lighting Engineers, Ltg. Res. & Technol., 14, 2, 1982.
[8] Littlefair, P.J. The Luminance Distribution of an Average Sky, Ltg. Res. & Technol., 13, 192, 1981.
[9] Page, J.K. EEC Solar Energy Program Project F – Solar Radiation Data. Final Report on the Predetermination of Radiation on Inclined Surfaces at Different European Centres, Department of Building and Science, Sheffield University, Sheffield, U.K.
[10] Aydinli, S. The Calculation of the Available Solar Radiation and Daylight, Proc. CIE Symposium on Daylight, Berlin, 1980.
[11] Littlefair, P.J. Designing for Daylight Availability using the BRE Average Sky, Proceedings of CIBS National Lighting Conference, Warwick, 1982.
[12] van Bergem-Jansen, P.M. and Soeleman, R.S. Window Design – Visual and Thermal Consequences, Proceedings, CIE Symposium on Daylight, Berlin, 1980.
[13] Hopkinson, R.G. Architectural Physics: Lighting, HMSO, London, 1963.

**Table A9.6**   Constant *a* for the internally reflected component

| Floor reflection factor | 0.3 | | | 0.1 | | | | | | | |
|---|---|---|---|---|---|---|---|---|---|---|---|
| Ceiling reflection factor | 0.7 | | | 0.7 | | | 0.5 | | | 0.3 | |
| Wall reflection factor | 0.5 | 0.3 | 0.1 | 0.5 | 0.3 | 0.1 | 0.5 | 0.3 | 0.1 | 0.3 | 0.1 |
| Room index | Values of *a* | | | | | | | | | | |
| 1.0 | 1.1 | 1.1 | 1.0 | 1.0 | 1.0 | 1.0 | 0.9 | 0.9 | 0.9 | 0.9 | 0.9 |
| 1.25 | 1.1 | 1.1 | 1.1 | 1.1 | 1.0 | 1.0 | 1.0 | 1.0 | 1.0 | 1.0 | 1.0 |
| 1.5 | 1.2 | 1.1 | 1.1 | 1.2 | 1.1 | 1.1 | 1.1 | 1.1 | 1.0 | 1.0 | 1.0 |
| 2.0 | 1.2 | 1.2 | 1.1 | 1.2 | 1.1 | 1.1 | 1.1 | 1.1 | 1.0 | 1.0 | 1.0 |
| 2.5 | 1.3 | 1.2 | 1.2 | 1.3 | 1.1 | 1.1 | 1.2 | 1.1 | 1.0 | 1.0 | 1.0 |
| 3.0 | 1.5 | 1.4 | 1.3 | 1.4 | 1.2 | 1.1 | 1.3 | 1.2 | 1.1 | 1.1 | 1.0 |
| 4.0 | 1.7 | 1.6 | 1.4 | 1.5 | 1.3 | 1.2 | 1.4 | 1.3 | 1.2 | 1.1 | 1.0 |
| 5.0 | 2.0 | 1.8 | 1.6 | 1.7 | 1.4 | 1.3 | 1.5 | 1.4 | 1.3 | 1.1 | 1.0 |

**Table A9.7**   Constant *v* for the internally reflected component

| Floor reflection factor | 0.3 | | | 0.1 | | | | | | | |
|---|---|---|---|---|---|---|---|---|---|---|---|
| Ceiling reflection factor | 0.7 | | | 0.7 | | | 0.5 | | | 0.3 | |
| Wall reflection factor | 0.5 | 0.3 | 0.1 | 0.5 | 0.3 | 0.1 | 0.5 | 0.3 | 0.1 | 0.3 | 0.1 |
| Room index | Values of *v* | | | | | | | | | | |
| 1.0 | 4.0 | 2.9 | 2.1 | 3.5 | 2.2 | 1.6 | 3.1 | 2.0 | 1.3 | 1.7 | 1.0 |
| 1.25 | 3.9 | 2.6 | 2.0 | 3.1 | 2.0 | 1.6 | 2.7 | 1.8 | 1.3 | 1.6 | 0.9 |
| 1.5 | 3.8 | 2.3 | 1.8 | 2.7 | 1.8 | 1.4 | 2.5 | 1.6 | 1.1 | 1.3 | 0.8 |
| 2.0 | 3.5 | 2.2 | 1.7 | 2.5 | 1.7 | 1.4 | 2.1 | 1.4 | 1.0 | 1.1 | 0.8 |
| 2.5 | 3.2 | 2.0 | 1.6 | 2.3 | 1.6 | 1.3 | 1.8 | 1.3 | 0.9 | 1.0 | 0.6 |
| 3.0 | 2.7 | 1.7 | 1.3 | 2.1 | 1.4 | 1.1 | 1.6 | 1.1 | 0.9 | 1.0 | 0.6 |
| 4.0 | 2.5 | 1.6 | 1.1 | 1.8 | 1.3 | 1.0 | 1.3 | 1.0 | 0.8 | 0.9 | 0.5 |
| 5.0 | 2.1 | 1.3 | 1.0 | 1.6 | 1.1 | 0.9 | 1.0 | 0.9 | 0.6 | 0.8 | 0.4 |

**Table A9.8**   Constant *e* for the internally reflected component

| Angle of obstruction from centre of window (degrees above horizontal) | *e* |
|---|---|
| 0° (i.e., unobstructed) | 1.0 |
| 10° | 0.9 |
| 20° | 0.8 |
| 30° | 0.65 |
| 40° | 0.5 |
| 50° | 0.35 |
| 60° | 0.25 |
| 70° | 0.18 |
| 80° | 0.13 |

**Table A9.9**   Orientation factors for use with overcast sky, daylight factors

| | Orientation factor |
|---|---|
| South-facing window | 1.20 |
| East-facing window | 1.04 |
| West-facing window | 1.00 |
| North-facing window | 0.77 |

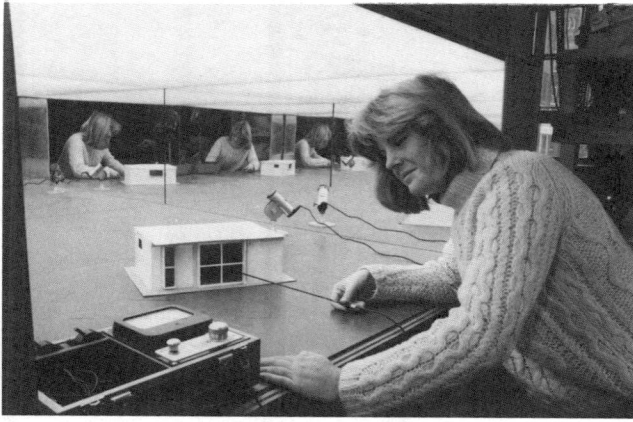

**Fig. A9.6** Measuring the daylight factor for a model inside an artificial sky

## APPENDIX 10  ILLUMINANCE AT A POINT REFERENCE DIAGRAMS

### A10.1  Introduction

On the following pages is a series of diagrams. Each diagram shows a particular arrangement of a light source illuminating a point on a surface. With each diagram is the associated formula necessary for the calculation of illuminance at the point.

In order to use the reference diagrams, first determine whether the source is a point source, a line source or an area source. To do this, calculate the distance between the centre of the luminaire and the point for which the illuminance is to be calculated. If this distance is $D$ and the width and length of the fitting are $W$ and $L$ respectively. Then

If $5W \leqslant D$, *and* $5L \leqslant D$, then use the point source formulae (Section A10.2).

If $5W \leqslant D$, *but* $5L > D$, then use the line source formulae (Section A10.3).

If $5W > D$, *and* $5L > D$, then use the area source formulae (Section A10.4).

Having established which set of formulae to use, turn to the appropriate section and read the notes, before looking through the individual diagrams to find one which matches your problem.

*Note:* Where luminous intensity values are obtained from published photometric data they will normally be quoted in candelas per 1000 lamp lumens. These should be corrected by multiplying by the total bare lamp luminous flux of the luminaire divided by 1000.

### A10.2  Point source formulae

Three applications of the inverse square and cosine laws are given: *(a)* the general case from which the others are derived (Fig A10.1); *(b)* the illuminance on a horizontal surface (Fig A10.2); *(c)* the illuminance on a vertical surface (Fig A10.3).

In each of the formulae the luminous intensity $I(\theta)$ in candelas at the angle of elevation $(\theta)$ is required. This can be found from the luminous intensity distribution of the luminaire.

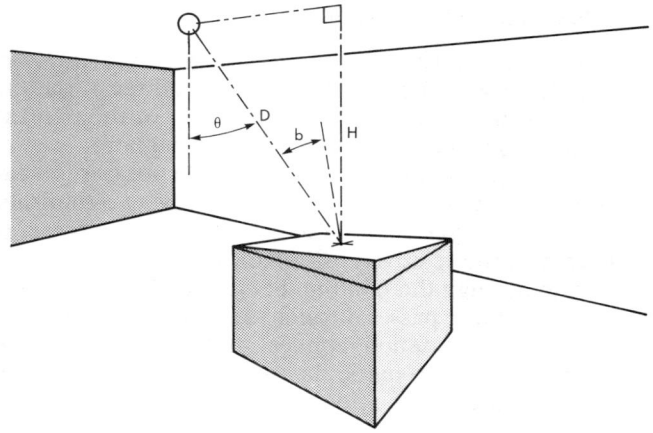

**Fig A10.1** Point source formulae general case – illuminance on a plane at an angle $b$ to the source

$$E = \frac{I(\theta).\cos b}{D^2}$$

or

$$E = \frac{I(\theta).\cos^2(\theta).\cos b}{H^2}$$

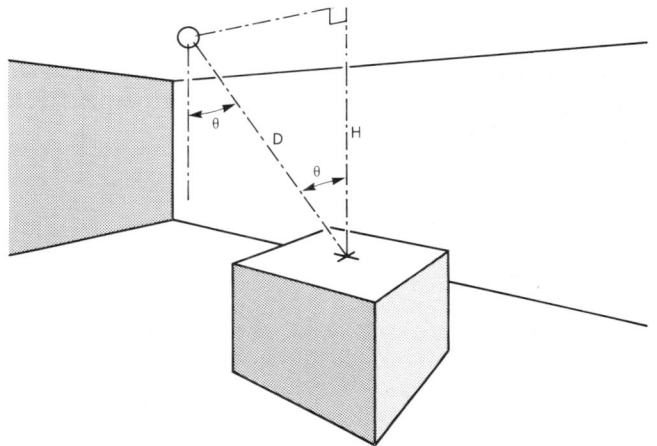

**Fig. A10.2** Point source formulae illuminance on a horizontal plane

$$E = \frac{I(\theta).\cos(\theta)}{D^2}$$

or

$$E = \frac{I(\theta).\cos^3(\theta)}{H^2}$$

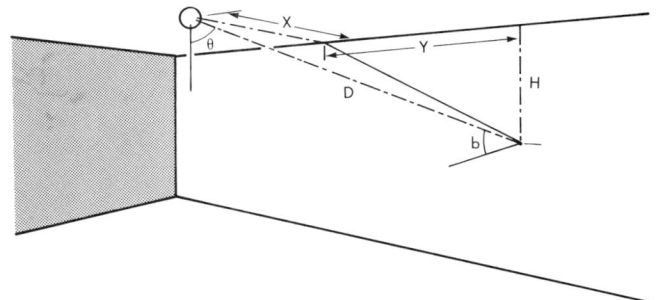

**Fig. A10.3** Point source formulae illuminance on a vertical plane

$$E = \frac{I(\theta).\cos b}{D^2} \quad or \quad E = \frac{I(\theta).\sin a}{D^2}$$

Where $\sin a = \dfrac{X}{D}$

## A10.3   Line source formulae

Line source formulae are given for four situations (labelled Fig A10.4 to Fig A10.7) with three variations for each case according to the position of the point of illumination relative to the end of the luminaire.

The most suitable case for any given application can be obtained by studying the figures. Having established which case applies to the situation, it is then necessary to determine the relative position of the source. If a plane passing through the point and perpendicular to the axis of the luminaire passes through the end of the luminaire, then the 'a' variation applies. If the plane cuts the luminaire into two parts, then the 'b' variation must be used, and if the plane does not cut the luminaire at all, then the 'c' situation applies.

The differences between these different cases are most readily seen by examining the different diagrams. In each formula, $I(\theta)$ is the luminous intensity of the luminaire at an angle $(\theta)$ in the *transverse* plane; i.e. in a plane at right angles to the luminaire axis. The angles $a_1$ and $a_2$ are called the 'aspect angles' and are used to obtain the aspect factors from aspect factor tables. In all cases:

$$a_1 = \tan^{-1}\left(\frac{S_1}{D}\right) \qquad a_2 = \tan^{-1}\left(\frac{S_2}{D}\right)$$

and $AF(a_1)$ denotes the parallel plane aspect factor for an angle of $a_1$, whilst $af(a_2)$ denotes a perpendicular plane aspect for an angle $a_2$.

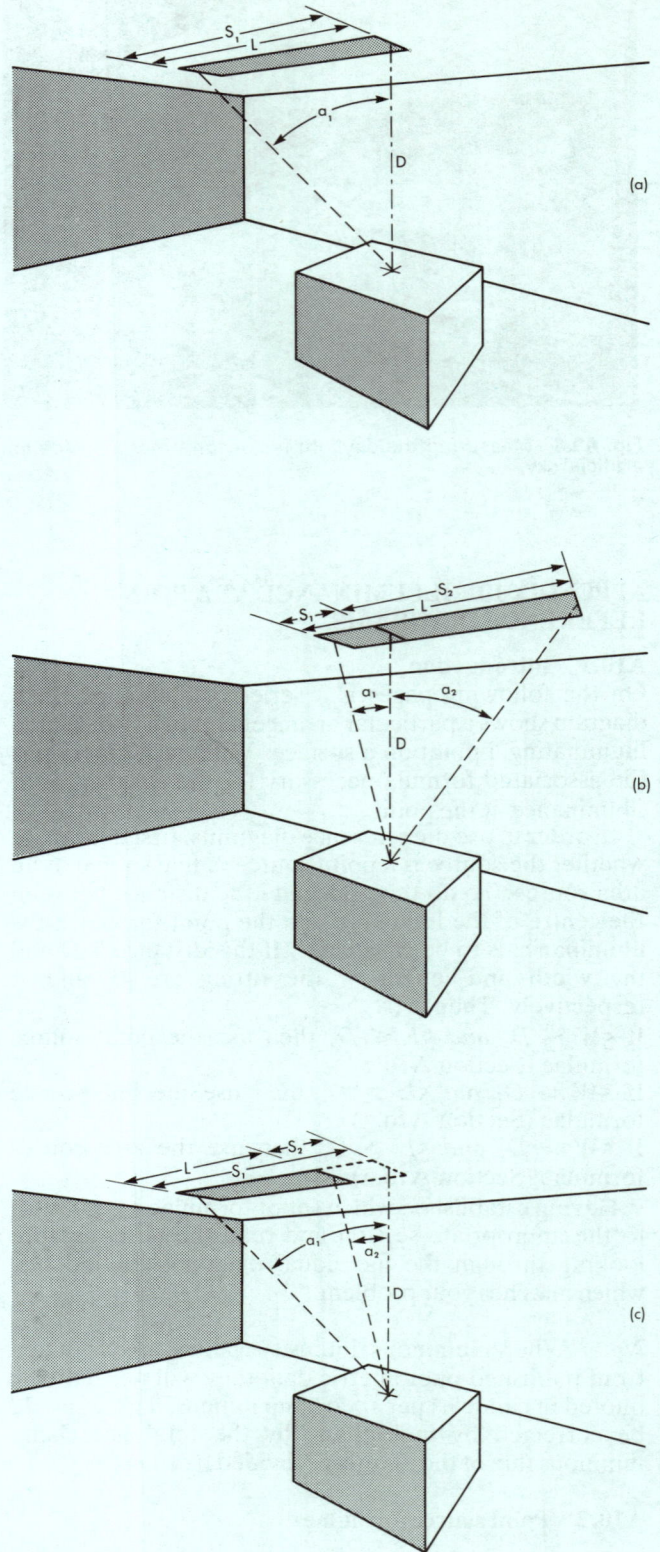

(a)

(b)

(c)

**Fig. A10.4**   Line source formulae, illuminance on a horizontal surface

a)
$$E = \frac{I(\theta).AF(a_1)}{L.D}$$

b)
$$E = \frac{I(\theta).(AF(a_1) + AF(a_2))}{L.D}$$

c)
$$E = \frac{I(\theta).(AF(a_1) - AF(a_2))}{L.D}$$

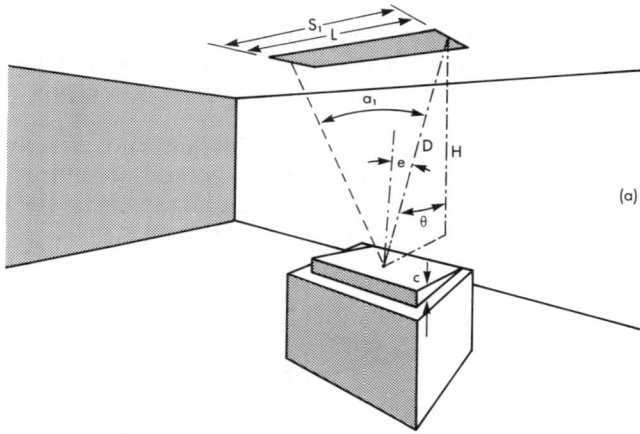

(a)

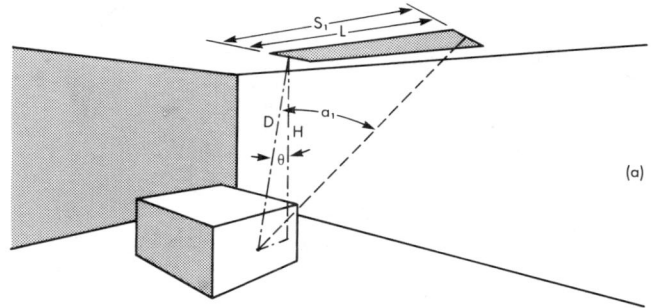

(a)

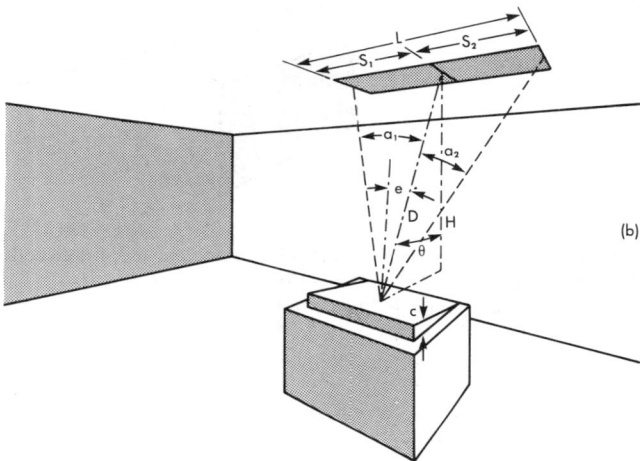

(b)

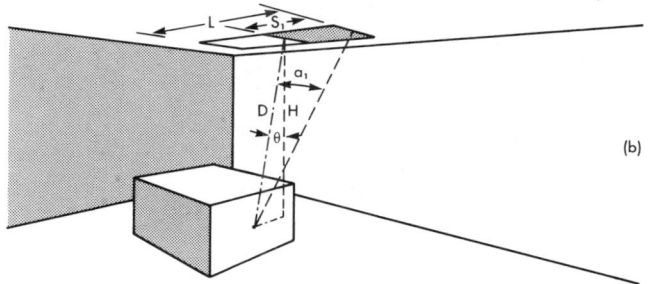

(b)

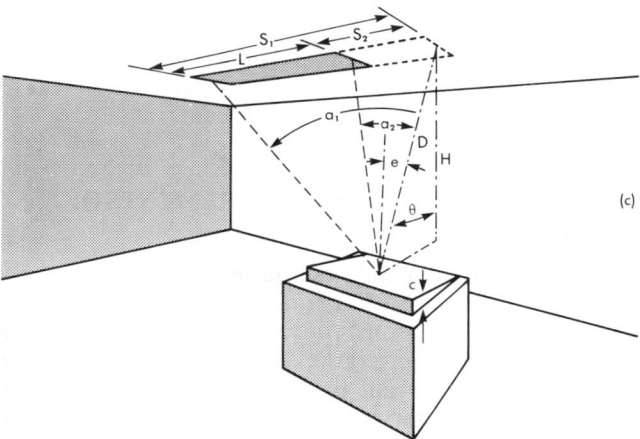

(c)

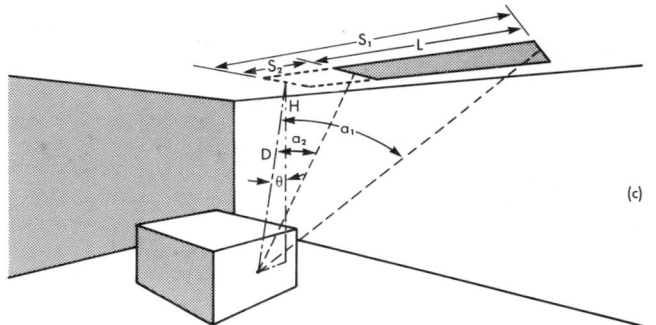

(c)

**Fig. A10.5** Line source formulae, illuminance on an inclined (or vertical) surface parallel to the axis of the luminaire $(e = \theta - c)$

a) $$E = \frac{I(\theta) \cdot \cos e \cdot AF(a_1)}{L.D}$$

b) $$E = \frac{I(\theta) \cdot \cos e \cdot (AF(a_1) + AF(a_2))}{L.D}$$

c) $$E = \frac{I(\theta) \cdot \cos e \cdot (AF(a_1) - AF(a_2))}{L.D}$$

**Fig. A10.6** Line source formulae, illuminance on a vertical surface perpendicular to the axis of the luminaire

a) $$E = \frac{I(\theta) \cdot af(a_1)}{L.D}$$

b) $$E = \frac{I(\theta) \cdot af(a_1)}{L.D}$$

c) $$E = \frac{I(\theta) \cdot (af(a_1) - af(a_2))}{L.D}$$

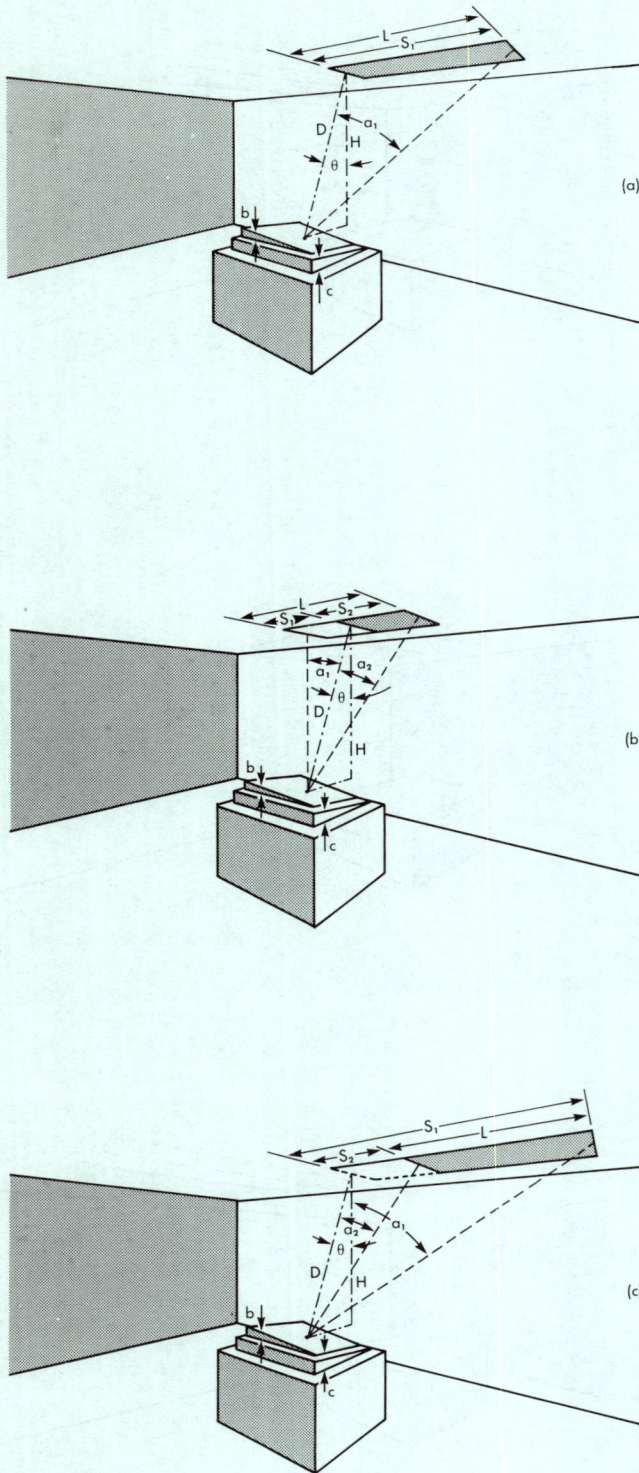

## A10.4  Area Source Formula

The basic formula for a uniform area source with a cosine distribution gives the illuminance at a point directly beneath one corner. The geometry is given in Fig A10.8.

To obtain the illuminance at a point that is not directly beneath one corner, it is necessary to add or subtract contributions from 4 imaginary area sources, each with a corner over the point, to obtain the resultant.

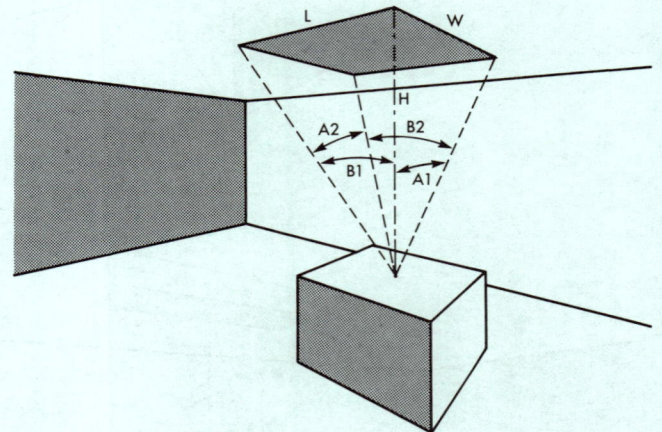

**Fig. A10.8**  Area source formula, illuminance at a point directly beneath one corner of the area source

$$E = \frac{Ip.(A2.\sin B1 + B2.\sin A1)}{2}$$

Where:

$A1 = \tan^{-1}(W/H),$

$A2 = \tan^{-1}(W/\sqrt{(L^2 + H^2)})$

$B1 = \tan^{-1}(L/H)$

$B2 = \tan^{-1}(L/\sqrt{(W^2 + H^2)})$

$Ip$ = Peak luminous intensity (It is assumed that $I(\theta) = Ip.\cos\theta$ )

## APPENDIX A11  FIELD MEASUREMENTS OF LIGHTING

### A11.1  Functions of field measurements

Field measurements of lighting are usually undertaken for one of three reasons: *(a)* to establish whether a new installation has achieved the design specification; *(b)* to establish whether an installation meets a desired criterion; *(c)* as part of a process for identifying the causes of complaints about the lighting, i.e. trouble-shooting.

The same instrumentation is used for all three purposes although the nature of the measurements made will vary with the circumstances.

### A11.2  Instruments

The vast majority of field measurements of lighting are undertaken with two basic instruments, an illuminance meter and a luminance meter.

**Fig. A10.7**  Line source formulae, general case, illuminance on any inclined surface ($e = \theta - c$)

a)
$$E = \frac{I(\theta).\cos e.\cos b. AF(a_1) + I(\theta).\sin b. af(a_1)}{L.D}$$

b)
$$E = \frac{I(\theta).\cos e.\cos b. (AF(a_1) + AF(a_2)) + I(\theta).\sin b.(af(a_1) + af(a_2))}{L.D}$$

c)
$$E = \frac{I(\theta).\cos e.\cos b. (AF(a_1) - AF(a_2)) + I(\theta).\sin b.(af(a_1) - af(a_2))}{L.D}$$

*Appendix 10 Illuminance at a point reference diagrams*

## A11.2.1 Illuminance meters

Illuminance meters usually consist of a selenium or silicon photovoltaic cell connected directly or indirectly by an amplifier to a display which can be analogue or digital. The quality of an illuminance meter is determined by four factors: *(a)* its spectral response; *(b)* its response to light incident on the photocell at different angles; *(c)* its linearity of response; and *(d)* its sensitivity to temperature.

The basic spectral response of both selenium and silicon photovoltaic cells differs from that of the human visual system. Therefore to achieve accurate measurements of illuminance it is necessary to correct the spectral response of the photocell to that of the human visual system. This can be done either directly, by means of a filter superimposed on the photocell, or indirectly by providing correction factors to effectively recalibrate the photocell for different light sources. When filters are used the instrument is described as colour corrected. The photocell whose spectral sensitivity is corrected by a filter can be used for all light sources, either separately or in combination, although the accuracy of the result will obviously depend on the quality of the filter. The photocell whose spectral sensitivity is modified by correction factors supplied by the manufacturers, can only be used for those light sources for which correction factors are available and then only for those light sources when they occur alone.

The response of illuminance meters to light falling on the photocell from different directions, is termed the response to oblique light incidence (or cosine response). Specifically, the measured illuminance $E$ for light incident at an angle $\theta$ from the normal $n$ to the photocell should follow the equation $E = E_n \cos \theta$. Illuminance meters which are not cosine corrected can give large measurement errors when used to measure illuminances where an appreciable proportion of the luminous flux comes at large deviations from the normal, e.g. when measuring daylight in side-lit rooms. Most illuminance meters are cosine corrected by means of either transparent hemispheres or diffusing covers of some sort. It is important that these covers are kept clean.

The linearity of response of an illuminance meter is determined by the resistance of the circuit into which the output from the photocell is fed; the higher the resistance the greater will be the non-linearity of response at higher illuminances.

The sensitivity of illuminance meters to temperature variations is also influenced by the resistance of the circuitry associated with the photocell. If that resistance is high then extremes of temperatures will cause errors in measurement. Selenium photocells are considerably more sensitive to temperature than are silicon photocells. Prolonged exposure to temperatures above 50°C will permanently damage selenium photocells. Ideally photovoltaic cells should be operated in ambient temperatures of about 25°C. They should not be used outside the temperature range 15°C to 50°C. Within this temperature range errors will still occur, but correction factors for different operating temperatures can be supplied by manufacturers.

To summarise, a good illuminance meter should be colour and cosine corrected, should be linear in response and insensitive to ambient temperature variations. Standards for two grades of portable photoelectric illuminance meters (Types P1 and P2) are given in BS 667[1]. Errors of measurement of ±10% (Type 1) and ±15% (Type 2) are permitted. This gives some idea of what is achievable even with a good quality illuminance meter when it is new. It should also be noted that the sensitivity of illuminance meters varies with time. Illuminance meters should be recalibrated at least once a year. This can be done by any photometric laboratory. Illuminance meters are available for illuminances from 0.1 lux to 100,000 lux full scale deflection i.e. from emergency lighting conditions to daylight conditions.

## A11.2.2 Luminance meters

A luminance meter consists of an imaging system, a photo-receptor, and a display. The optical imaging system is used to form an image of the object of interest on the photo-receptor. The photo-receptor produces a signal which is dependent on the luminance of the part of the object of interest it receives. This signal is amplified and displayed in either analogue or digital form. By changing the imaging system it is possible to alter the field of view of the photo-receptor and hence to have different areas of measurement. The photo-receptors used in luminance meters may be photovoltaic cells or a photomultiplier tube. The photovoltaic cells used in luminance meters are the same as those used in illuminance meters. When used in luminance meters the photovoltaic cells again need to have their spectral sensitivity corrected to that of the human visual system and to be combined with the associated circuitry in such a way as to have a linear response. They will also be sensitive to ambient temperature conditions.

Photomultiplier tubes are more sensitive to luminous flux than photovoltaic cells. They are also more sensitive to vibration and magnetic fields and need a high voltage electrical supply. Like photovoltaic cells photomultiplier tubes need to have their spectral response corrected to match the spectral response of the human visual system. The choice between photovoltaic cells and photomultiplier tubes is essentially one of the balance required between sensitivity and robustness. Photomultiplier tubes are more sensitive but are also less than robust.

Luminance meters which provide measurements over a range of $10^{-4}$ to $10^8$ cd/m$^2$ for areas varying from a few seconds of arc to several degrees are available.

## A11.3 Fields surveys

### A11.3.1 General

Field measurements are always obtained in specific circumstances. It is therefore essential when making field measurements to keep a complete and accurate record of the state of the lighting installation and the interior in general at the time the measurements are made. Particular attention should be given to the lamp type and age, the level and stability of the supply voltage, the state of maintenance of the lamps and luminaires, the surface reflectances, the degree of obstruction and any other factors which may be expected to influence the measurement. Photographs of the interior are a valuable supplement to a written record.

Before starting measurements it is necessary to decide on the conditions of interest. For example, is daylight to be admitted and if it is what type of control is to be used; are the measurements to be concerned with average conditions over the interior or are they concerned only with individual work places; should the measurements around the workplace be taken with the people present etc. The answers to these and similar questions are determined by the aim of the survey.

In addition, before starting measurements it is necessary to stabilise the performance of the lamps and luminaires and of the illuminance and luminance meters used. The time required to stabilise the light output of an installation depends on the nature of the lamp and luminaire. Installations using discharge lamps, including tubular fluorescents, should be lit for at least 20 minutes and ideally for 1 hour before measurements are made. Installations using incandescent lamps should be lit for at least 10 minutes before any measurements are made.

It should be noted that daylight is rarely stable and hence the illuminance and luminance it produces can vary over a very large range very quickly. For this reason when measurements of the electric lighting installation alone are required, daylight must be excluded from the interior.

To stablise the reading of the photovoltaic cells used in illuminance and luminance meters it is desirable to expose the photocell to the approximate luminous flux to be measured for about 5 minutes before making the first measurement.

### A11.3.2 Average illuminance

The average illuminance over an interior is usually measured to check if an electric lighting installation has achieved its design specification. To do this the following procedure is recommended, after the installation has been operating for an appropriate time at the design supply voltage. For discharge lamps this time is 100 hours but for incandescent lamps it will be less.

The interior is divided into a number of equal areas which should be as nearly square as possible. The illuminance at the centre of each area is measured and the mean value calculated. This gives an estimate of the average illuminance. The accuracy of the estimate depends on the number of measurement points and the uniformity of illuminance.

Table A11.1 relates the room index to the number of measurement points necessary to give an error of less than 10 per cent; the data in the Table are valid for spacing/height ratios up to 1.5:1. Where an error of less than 5 per cent is required, the number of measurement points should be doubled.

The only limitation on the use of Table A11.1 is when the periodicity of the grid of measuring points coincides with the periodicity of the grid of lighting points; large errors are then possible and more points than the number given in Table A11.1 should be used. The numbers of measurement points suggested are minima, and it may be necessary to increase their number to obtain a symmetrical grid to suit a particular room shape.

**Table A11.1** Relationship between room index and the minimum number of measurement points

| Room index | Number of points |
|---|---|
| Below 1 | 4 |
| 1 and below 2 | 9 |
| 2 and below 3 | 16 |
| 3 and above | 25 |

The following examples illustrate the use of the method:

(a) For an interior measuring 20 m × 20 m and with luminaires mounted 4 m above the working plane

$$\text{Room index} = \frac{20 \times 20}{4(20 + 20)} = 2.5$$

Sixteen points of measurement are therefore required, i.e. a 4 × 4 grid.

(b) If the room measures 20 m × 40 m and the luminaires are mounted at the same height, it should be treated as two 20 m × 20 m areas and thirty-two points of measurement should be used.

(c) If the room measures 20 m × 33 m, the number of measurement points required should be derived by first considering a 20 m × 20 m area within the larger rectangle. From example (a) and treating this area by itself, sixteen points would be required. The number for the room is then obtained proportionately, i.e.

$$\text{Number of points} = 16 \times \left( \frac{20 \times 33}{20 \times 20} \right) = 26$$

The points are placed at the centres of rectangles which should be as 'square' as possible. Taking twenty-six as the minimum number of points, twenty-eight points on a 4 × 7 grid could be used.

Measurements should be made at a position representative of the working plane but if this is not specified the measurements should be taken on a horizontal plane at height of 0.7 m above the floor for offices and of 0.85 m above the floor for industrial premises. A portable stand or tripod is useful to support the photocell at the required height and inclination. Care should be taken not to cast a shadow over the photocell when taking the readings.

### A11.3.3 Illuminance at a point

When the illuminance at a workplace is of interest, e.g. when local lighting is being measured, the illuminance should be measured at an appropriate point and plane with the worker in his normal position, no matter whether this casts a shadow on the meter or not.

Point illuminance measurements can also be used to estimate the uniformity of the illuminance provided by the electric lighting installation. For this purpose the illuminance should be measured on a horizontal plane at an appropriate height without shadowing the photocell. Uniformity criteria need to be interpreted with a degree of common sense. It is always possible to find a very low illuminance in the corner of a room but this is of little relevance if no work is done there. It is the uniformity of the illuminance over the working area that usually is of concern.

The illuminance at a point is usually measured with an illuminance meter. However if no illuminance meter is available but a luminance meter is, the illuminance at a point can be calculated from measurements of the luminance of a matt surface of known reflectance.

### A11.3.4  Scalar illuminance

Scalar illuminance is a useful measure of the lighting conditions when there are no obvious working planes in the interior (see Appendix 1). For precise measurements of scalar illuminance, instruments based on a photocell with a diffusing sphere attached are available. However, a good approximation to the scalar illuminance at a point can be obtained using a conventional illuminance meter by averaging the illuminances measured on the six faces of a cube centred at the point. An even better approximation, particularly where there is little inter-reflected light, is given by the average of the illuminances on four faces of a regular tetrahedron centred at the point.

### A11.3.5  Mean cylindrical illuminance

Mean cylindrical illuminance is a useful measure of the lighting in interiors where vertical planes are important (see Appendix 1). For precise measurements of mean cylindrical illuminance instruments based on a photocell with a uniform diffusing cylinder attachment are available. However, a reasonable approximation to mean cylindrical illuminance can be obtained using a conventional illuminance meter from the average of the illuminances on the four vertical surfaces of a cube. An even better approximation, particularly where there is little inter-reflected light available, is given by the average of the illuminances on the three vertical faces of a rod of a 120° triangular section.

### A11.3.6  Illumination vector

The illumination vector is a measure of the 'flow of light' in a room. It can be combined with the scalar illuminance to give the vector/scalar ratio which is a measure of the modelling effect of the lighting. Precise measurements of the magnitude and direction of the illumination vector can be made using commercially available equipment based on a matched pair of cosine corrected photovoltaic cells mounted back-to-back Alternatively, measurements of magnitude and direction can be obtained by the vector addition of the differences in illuminance between the three pairs of opposing faces of a cube centred at the point. Yet another approach is to measure the two components of the illumination vector separately. The vector direction can be identified by the use of a pair of grease spot photometers mounted at right angles[2]. Once the direction of the illumination vector at a point has been established, the magnitude of the illumination vector can be obtained from the difference in illuminances on the opposing faces of the plane through which the vector direction passes normally.

### A11.3.7  Luminance measurements

Luminance measurements are most frequently made in trouble-shooting investigations, particulary when the complaints are of over-brightness. In these circumstances the conditions which are the subject of complaint should be established and luminance measurements made from the position of the people who are complaining. In this way the source of the complaints may be identified. Luminances are most accurately and conveniently measured with a luminance meter. However, where one is not available, an estimate of the luminance of a surface can be obtained indirectly by measuring the reflectance of the surface and the illuminance on it and then calculating the luminance.

### A11.3.8  Measurement of reflectance

The reflectance (and luminance factor) of a surface can be most accurately measured by the use of a luminance meter and a standard reflectance surface made from pressed barium sulphate or magnesium oxide. The luminances of the surface of interest and the standard reflectance surface are measured from an appropriate position. Then the reflectance (or luminance factor) of the surface of interest is given by the expression

$R = R_s. L_1/L_s,$

where $R$ = reflectance (or luminance factor) of the surface of interest, $L_1$ = luminance of the surface of interest cd/m$^2$, $L_s$ = luminance of the standard reflectance surface cd/m$^2$, $R_s$ = reflectance (or luminance factor) of the standard reflectance surface.

If a luminance meter is not available, then an approximate measure of the reflectance of a surface can be obtained by making a match between the surface of interest and a sample from a range of colour samples of known reflectance (see Appendix 3).

### A11.3.9  Daylight measurements

As stated earlier, the illuminance and luminance produced in an interior by daylight can vary widely and rapidly. For this reason, the daylight available in an interior is usually expressed in terms of daylight factor rather than in terms of illuminance and luminance. The latter measures, although they can be obtained, are not very meaningful. Daylight factor can be measured. Full details are given in BS DD73 and will be given in a forthcoming CIBS publication[3,4].

## A11.4  References

[1] British Standards Institution, BS 667, Portable Photo-electric Photometers. BSI London.
[2] Dale, N.P.G., Broadbridge, J.B. & Crowther, P.M. Measuring the direction of the flow of light. Ltg. Res. & Technol. **4**, 43, 1972.
[3] British Standards Institution BS DD73 1982 – Basic data for the design of buildings: Daylight, BSI, London, 1982.
[4] Chartered Institution of Building Services, Window Design Guide, to be published.

# GLOSSARY

The explanations and definitions given in this Glossary are intended to help readers to understand the Code. They are based on British Standard 4727: Part 4: Glossary of terms particular to lighting and colour, 1971/2, and on the 3rd edition of the International Lighting Vocabulary issued jointly by the Commission Internationale de l'Eclairage (CIE) and the International Electrotechnical Commission (IEC). These documents should be consulted if more precise definitions are required.

## adaptation
The process which takes place as the visual system adjusts itself to the brightness or the colour of the visual field. The term is also used, usually qualified, to denote the final state of this process. For example 'dark adaptation' denotes the state of the visual system when it has become adapted to a very low luminance.

## apostilb (asb)
A non-SI unit of luminance. One apostilb is the luminance of a uniform diffuser emitting one lumen per square metre (see conversion table for conversion to SI units of luminance).

## apparent colour
Of a light source; subjectively the hue of the source or of a white surface illuminated by the source; the degree of warmth associated with the source colour. Lamps of low correlated colour temperatures are usually described as having a warm apparent colour, and lamps of high correlated colour temperature as having a cold apparent colour.

## aspect factor (AF)
A function of the angle subtended at a point by the length of a linear source, and of the axial distribution of luminous intensity from the source; used in the calculation of illuminance at a point.

## BZ (British Zonal) System
A system for classifying luminaires as described in CIBS Technical Memorandum No. 5. The BZ class number (e.g. BZ 3) denotes the classification of a luminaire in terms of the flux from a conventional installation directly incident on the working plane, relative to the total flux emitted below the horizontal (the direct radio).

## candela (cd)
The SI unit of luminous intensity, equal to one lumen per steradian.

## cavity index (CI)
A term, indicating the proportions of boundary surfaces, used in determining the effective reflectances of room surfaces for interior lighting design: defined for a cavity of length $L$, width $W$, and depth $d$, as $LW/(d(L + W))$.

## ceiling cavity reflectance (RE(C))
Effective reflectance of the room volume above the plane of the luminaires.

## chroma
In the Munsell system, an index of saturation of colour ranging from 0 for neutral grey to 16 for strong colours. A low chroma implies a pastel shade.

## chromaticity
The colour quality of a stimulus, usually defined by coordinates on a plane diagram in the CIE colorimetric system (CIE Publication 15) or by the combination of dominant wavelength and purity.

## CIE chromaticity diagram
A plane diagram showing the effect of mixtures of colour stimuli, each chromaticity being represented unambiguously by a single point on the diagram.

## CIE Standard Photometric observer
A receptor of radiation whose relative spectral sensitivity curve conforms to the $V(\lambda)$ curve or to the $V^I(\lambda)$. (CIE Publication 41.)

## colour atlas
A collection of colour samples arranged according to specified rules.

## colour rendering
A general expression for the appearance of surface colours when illuminated by light from a given source compared, consciously or unconsciously, with their appearance under light from some reference source. 'Good colour rendering' implies similarity of appearance to that under an acceptable light source, such as daylight.

## colour rendering index (CRI)
A measure of the degree to which the colours of surfaces illuminated by a given light source conform to those of the same surfaces under a reference illuminant, suitable allowance having been made for the state of chromatic adaptation. (CIE Publication 13.2).

## colour solid
That part of a colour space that is occupied by surface colours.

## colour space
A geometric representation of colours in space, usually of three dimensions.

## colour temperature (K)
The temperature of a full radiator which emits radiation of the same chromaticity as the radiator being considered.

## contrast
A term that is used subjectively and objectively. Subjectively it describes the difference in appearance of two parts of a visual field seen simultaneously or successively. The difference may be one of brightness or colour or both. Objectively, the term expresses the luminance difference between the two parts of the field by such relationships as:

$$\text{contrast} = \left| \frac{L - L_1}{L_1} \right|$$

Quantitatively, the sign of the contrast is ignored. $L_1$ is the dominant or background luminance. $L$ is the task luminance.

*contrast rendering factor (CRF)*
The ratio of the contrast of a task under a given lighting installation to its contrast under reference lighting conditions.

*contrast sensitivity*
The reciprocal of the minimum perceptible contrast.

*correlated colour temperature (unit:* K)
The temperature of a full radiator which emits radiation having a chromaticity nearest to that of the light source being considered, e.g., the colour of a full radiator at 3500K is the nearest match to that of a White tubular flourescent lamp.

*daylight factor*
The illuminance received at a point indoors, from a sky of known or assumed luminance distribution, expressed as a percentage of the horizontal illuminance outdoors from an unobstructed hemisphere of the same sky. Direct sunlight is excluded from both values of illuminance.

*design service illuminance*
The service illuminance used in the lighting specification. Design service illuminance is derived from the standard service illuminance by taking account of the modifying factors contained in the flow chart (Table 2.2).

*diffuse reflection*
Reflection in which the reflected light is diffused and there is no significant specular reflection, as from a matt paint.

*diffused lighting*
Lighting in which the luminous flux comes from many directions, none of which predominates.

*direct lighting*
Lighting in which the greater part of the luminous flux from the luminaires reaches the surface (usually the working plane) directly, i.e without reflection from surrounding surfaces. Luminaires with a flux fraction ratio less than 0.1 are usually regarded as direct.

*direct ratio (DR)*
The proportion of the total downward luminous flux from a conventional installation of luminaires which is directly incident on the working plane.

*directional lighting*
Lighting designed to illuminate a task or surface predominantly from some direction.

*disability glare*
Glare which impairs the ability to see detail.

*discharge lamp*
A lamp in which the light is produced either directly or by the excitation of phosphors by an electric discharge through a gas, a metal vapour or a mixture of several gases and vapours.

*discomfort glare*
Glare which causes visual discomfort.

*dominant wavelength*
The wavelength of a monochromatic light stimulus which, combined with an achromatic stimulus, gives a colour match with the stimulus being considered.

*downlighters*
Luminaires from which light is emitted only within relatively small angles to the downward vertical.

*downward light output ratio (DLOR)*
The ratio of the total light output of a luminaire below the horizontal under stated practical conditions to that of the lamp or lamps under reference conditions.

*effective reflectance (RE)*
Estimated reflectance of a surface, based on the relative areas and the reflectances of the materials forming the surface. Thus, 'effective wall reflectance' takes account of the reflectances of the wall surface, the windows, the filing cabinets, etc., that comprise the sides of a room.

*emergency lighting*
Lighting provided for use when the main lighting installation fails.

*escape lighting*
Emergency lighting provided to ensure that the means of escape can be safely and effectively used at all material times.

*externally reflected component of the daylight factor (ERC)*
The illuminance received directly at a point indoors from a sky of known or assumed luminance distribution after reflection from an external reflecting surface, expressed as a percentage of the horizontal illuminance outdoors from an unobstructed hemisphere of the same sky. Direct sunlight is excluded from both illuminances.

*flicker*
A visible oscillation in luminous flux.

*floor cavity reflectance (RE(F))*
Effective reflectance of the room volume below the working plane.

*flux fraction*
The proportion of luminous flux emitted from a luminaire in the upper or lower hemisphere (upper and lower flux fractions).

*flux fraction ratio (FFR)*
The ratio of the upward luminous flux to the downward luminous flux from a luminaire. It is also the ratio of the upper flux fraction to the lower flux fraction and the ratio of the upward light output ratio to the downward light output ratio.

*foot-candle (fc)*
A non-SI unit of illuminance, until recently used in the

USA, having the same value as the lumen per square foot (1 foot-candle = 10.76 lumens/square metre).

*foot lambert (fL)*
A non-SI unit of luminance based on the same concept as the apostilb. One foot lambert is the luminance of a uniform diffuser emitting one lumen per square foot (see conversion table for conversions to the SI unit of luminance).

*full radiator*
A thermal radiator obeying Planck's radiation law and having the maximum possible radiant exitance for all wavelengths for a given temperature; also called a black body to emphasise its absorption of all incident radiation.

*full radiator locus*
The curve on a chromaticity diagram representing the colour of the radiation from a full radiator as a function of its temperature.

*general lighting*
Lighting designed to illuminate the whole of an area uniformly, without provision for special local requirements.

*general surround lighting*
Lighting designed to illuminate the non-working parts of a working interior.

*glare*
The discomfort or impairment of vision experienced when parts of the visual field are excessively bright in relation to the general surroundings.

*glare index system*
A system which produces a numerical index calculated according to the method described in forthcoming CIBS Technical Memorandum 10. It enables the discomfort glare from lighting installations to be ranked in order of severity and the permissible limit of discomfort glare from an installation to be prescribed quantitatively.

*greyness*
The estimated grey content of surface colours on a scale from maximum greyness to zero greyness; (from BS 5252).

*hazardous environment*
An environment in which a risk of fire or explosion exists.

*hostile environment*
An environment in which the lighting equipment may be subject to chemical, thermal or mechanical attack.

*hue*
Colour in the sense of red, or yellow or green, etc. In the Munsell system, an index derived by arranging the five named colours (red, yellow, green, blue and purple) and their intermediates (yellow-red, green-yellow, etc.) in a circle of constant chroma and dividing each of these ten equally spaced bands into ten equal steps in the complete atlas (but four in the standard atlases, hence 7.5 BG as a hue reference).

*ingress protection (IP) number*
A two-digit number associated with a luminaire. The first digit classifies the degree of protection the luminaire provides against the ingress of solid foreign bodies. The second digit classifies the degree of protection the luminaire provides against the ingress of moisture. Details of the nature of the protection achieved at different levels is given in BS 4533.

*illuminance (E) (unit: $lm/m^2$, lux)*
The luminous flux density at a surface, i.e. the luminous flux incident per unit area. (This quantity was formerly known as the illumination value or illumination level.)

*illumination*
The process of lighting.

*illumination vector ($\vec{E}$) (unit: lx)*
A term used to describe the directional characteristics of light at a point. Its magnitude is the difference in the illuminances on opposite sides of a flat surface which is so orientated that this difference is a maximum. Its direction is normal to this surface; the positive direction of the vector is from the higher illuminance to the lower illuminance.

*incandescent lamp*
A lamp in which light is produced by a filament heated to incandescence by the passage of an electric current.

*indirect lighting*
Lighting in which the greater part of the flux reaches the surface (usually the working plane) only after reflection at other surfaces and particularly at the roof or ceiling. Luminaires with a flux fraction ratio greater than 10 are usually regarded as indirect.

*initial light output (unit: lm)*
The luminous flux from a lamp after 100 hours of operation.

*installed efficacy (unit: lm/W)*
A factor which quantifies the efficiency of a lighting installation in converting electrical power to light. Specifically it is the product of the lamp circuit luminous efficacy and the utilization factor.

*internally reflected component of the daylight factor(IRC)*
The illuminance received at a point indoors from a sky of known or assumed luminance distribution after reflection within the interior, expressed as a percentage of the horizontal illuminance outdoors from an unobstructed hemisphere of the same sky. Direct sunlight is excluded from both illuminances.

*irradiance (unit: $W/m^2$)*
The radiant flux density at a surface, i.e. the radiant flux incident per unit area of the surface.

*isolux diagram*
A diagram showing contours of equal illuminance.

*lambert*
A non-SI unit of luminance based on the same concept as the apostilb; one lambert is the luminance of a

uniform diffuser emitting one lumen per square centimetre (see conversion table for conversion to the SI unit of luminance).

*lamp lumen maintenance factor (LLMF)*
The proportion of the initial light output of a lamp that is produced after a set time.

*light loss factor (LLF)*
The ratio of the illuminance provided by the installation at some stated time, with respect to the initial illuminance, i.e. that after 100 hours of operation. The light loss factor is the product of the lamp lumen maintenance factor, the luminaire maintenance factor and the room surface maintenance factor.

*light output ratio (LOR)*
The ratio of the total light output of a luminaire under stated practical conditions to that of the lamp or lamps under reference conditions. For the luminaire, the output is usually measured in the designated operating position at 25°C ambient temperature with control gear of the type usually supplied in a luminaire and operated at its normal voltage. For the lamp the output is measured at 25°C ambient temperature and with control gear of standard properties. This is a practical basis for evaluating the total light output to be expected under service conditions.

*lighting design lumens (LDL) (unit: lm)*
Lamps vary in flux output, both between themselves and through their operating lives. The lighting design lumen is a nominal value which is representative of the average light output of each type or size of lamp throughout its life.

*lightness*
A subjective estimate of the proportion of light diffusely reflected by a body.

*limiting glare index*
The maximum value of the Glare Index which is recommended for a specific lighting application.

*load factor*
The ratio of the energy actually consumed by a lighting installation over a specified period of time to the energy that would have been consumed had the lighting installation always been operating during the period of time.

*local lighting*
Lighting designed to illuminate a particular small area which usually does not extend far beyond the visual task, e.g. a desk light.

*localised lighting*
Lighting designed to illuminate an interior and at the same time to provide higher illuminances over a particular part or parts of the interior.

*lumen (lm)*
The SI unit of luminous flux, used in describing a quantity of light emitted by a source or received by a surface. A small source which has a uniform luminous intensity of one candela emits a total of $4\pi$ lumens in all directions and emits one lumen within unit solid angle.

*luminaire*
An apparatus which controls the distribution of light given by a lamp or lamps and which includes all the components necessary for fixing and protecting the lamps and for connecting them to the supply circuit. Luminaire has superseded the term lighting fitting.

*luminaire maintenance factor LMF*
The lumen output from a luminaire declines with time because of dirt deposition on and in the luminaire. The luminaire maintenance factor quantifies this decline, being the proportion of the initial light output from the luminaire that occurs after a set time, allowance having been made for the decline in light output from the lamp.

*luminance (L) (unit:* cd/m$^2$)
The physical measure of the stimulus which produces the sensation of brightness measured by the luminous intensity of the light emitted or reflected in a given direction from a surface element, divided by the area of the element in the same direction. The SI unit of luminance is the candela per square metre, the relationship between luminance and illuminance is given by the equation

$$\text{Luminance} = \frac{\text{Illuminance} \times \text{reflectance factor}}{\pi}$$

This equation applies to a matt surface. For a non matt surface, the reflectance is replaced by the luminance factor.

*luminance coefficient*
The ratio of the luminance of an element of surface to the illuminance on it for given angles of viewing and incident light.

*luminance factor (B)*
The ratio of the luminance of a reflecting surface, viewed in a given direction, to that of a perfect white uniform diffusing surface identically illuminated.

For a non-matt surface the luminance factor may be greater or less than the reflectance.

*luminous efficacy (unit:* lm/W)
The ratio of the luminous flux emitted by a lamp to the power consumed by the lamp. When the power consumed by control gear is taken into account this term is sometimes known as lamp circuit luminous efficacy and is expressed in lumens/circuit watt.

*luminous efficiency*
The ratio of the radiant flux weighted according to the CIE Standard Photometric Observer to the corresponding radiant flux.

*luminous flux (unit:* lm)
The light emitted by a source, or received by a surface. The quantity is derived from radiant flux by evaluating the radiation in accordance with the spectral sensitivity of the standard eye as described by the CIE Standard Photometric Observer.

*luminous intensity (unit:* cd)
A quantity which describes the power of a source or illuminated surface to emit light in a given direction. It is the luminous flux emitted in a very narrow cone containing the given direction divided by the solid angle of the cone: the result is expressed in candelas.

*luminous intensity distribution*
The distribution of the luminous intensity of a lamp or luminaire in all spatial directions. Luminous intensity distributions are usually shown in the form of a polar diagram or a table for a single vertical plane, in terms of candelas per 1000 lumens of lamp luminous flux.

*lux* (lx)
The SI unit of illuminance, equal to one lumen per square metre.

*maintenance factor (MF)*
The ratio of the illuminance provided by an installation in the average condition of dirtiness expected in service, to the illuminance from the same installation when clean. The maintenance factor is always less than unity.

*mean cylindrical illuminance (unit:* $lm/m^2$)
The average illuminance over the curved surface of a very small cylinder located at a given point. Unless otherwise stated, the axis of the cylinder is taken to be vertical.

*metamerism*
The phenomenon occurring when coloured objects which match under one illuminant do not match under another (object metamerism) or when illuminants of the same apparent colour do not have the same colour rendering properties (illuminant metamerism).

*mixed reflection*
Partly specular and partly diffused reflection, as from a smooth, glossy paint.

*mounting height* ($h_m$)
Usually the vertical distance between a luminaire and the working plane, but sometimes the distance between the luminaire and the floor.

*Munsell system*
A system of surface colour classification using uniform colour scales of hue, value and chroma. A typical Munsell designation of a colour is 7.5 BG6/2, where 7.5 BG (blue-green) is the hue reference, 6 is the value and 2 is the chroma reference number.

*operating efficacy (unit:* lm/W)
A term which quantifies the efficacy of a lighting installation in use. Specifically operating efficacy is the quotient of the installed efficacy of the installation and the load factor.

*optical radiation*
That part of the electro-magnetic spectrum from 100 nm to 1 mm.

*power factor*
In an electric circuit, the power factor is equal to the ratio of the root mean square power in watts to the product of the root mean square values of voltage and current; for sinusoidal wave forms the power factor is also equal to the cosine of the angle of phase difference between voltage and current.

*purity*
A measure of the proportions of the amounts of the mono-chromatic and specified achromatic light stimuli that, when additively mixed, match the colour stimulus. The proportions can be measured in different ways yielding either colorimetric purity or excitation purity.

*radiance (unit:* $W/m^2/sr$)
At a point on a surface, the quotient of the radiant intensity emitted from an element of the surface in a given direction by the area of the element in the same direction.

*radiant efficiency*
The ratio of the radiant flux to the power consumed.

*radiant exposure (unit:* $J/m^2$)
At a point on a surface, the product of the irradiance and its duration.

*radiant flux (unit:* W)
The power emitted, transferred or received as radiation.

*radiant intensity (unit:* W/sr)
Of a source in a given direction: the quotient of the radiant flux emitted in a narrow cone containing the direction, by the solid angle of that cone.

*reference lighting*
Perfectly diffuse and unpolarised lighting by CIE standard illuminant A.

*reflectance (R)*
The ratio of the luminous flux reflected from a surface to the luminous flux incident on it. Except for matt surfaces, reflectance depends on how the surface is illuminated but especially on the direction of the incident light and it spectral distribution. The value is always less than unity and is expressed as either a decimal or as a percentage.

*room index (RI)*
An index related to the dimensions of a room and used when calculating the utilization factor and other characteristics of the lighting installation;

$$room\ index = \frac{L.\ W}{h_m\ (L + W)}$$

where $L$ is the length of the room, $W$ the width and $h_m$ the height of the luminaires above the working plane.

*room surface maintenance factor (RSMF)*
The proportion of the illuminance provided by a lighting installation in a room after a set time compared with that which occurred when the room was clean, allowance having been made for the depreciation in lumen output of lamps and the effect of dirt deposition on luminaires.

saturation
The subjective estimate of the amount of pure chromatic colour present in a sample observed judged in proportion to its brightness.

scalar illuminance (unit: lm/m²)
The average illuminance over the whole surface of a very small sphere located at a given point.

service illuminance
The mean illuminance throughout the maintenance cycle of an installation, averaged over the relevant area. The area may be the whole of the working plane or just the area of the visual task and its immediate surround, depending on the lighting approach used.

sky component of the daylight factor (SC)
The illuminance received directly at a point indoors from a sky of known or assumed luminance distribution expressed as a percentage of the horizontal illuminance outdoors from an unobstructed hemisphere of the same sky. Direct sunlight is excluded from both values of illuminance.

solid angle (unit: sr)
The angle subtended by an area at a point and equal to the quotient of the projected area on a sphere, centred on the point, by the square of the radius of the sphere; expressed in steradians.

spacing/height ratio (SHR)
This ratio describes the distance between luminaire centres in relation to their height above the working plane. For a regular square arrangement of luminaires, it is the distance between adjacent luminaires divided by their height above the working plane. More generally,

$$\text{spacing/height ratio} = \frac{1}{h_m} \sqrt{\left(\frac{A}{N}\right)}$$

where $A$ is the total floor area, $N$ is the number of luminaires and $h_m$ is their height above the working plane.

specular reflection
Reflection without diffusion in accordance with the laws of optical reflection as in a mirror.

standard service illuminance
The service illuminance recommended for the assumed standard conditions of the application.

standby lighting
Emergency lighting provided to enable normal activities to continue.

steradian (sr)
The unit of solid angle. A complete sphere subtends $4\pi$ steradians from the centre.

stilb
A non-SI unit of luminance equal to one candela per square centimetre (see conversion table for conversion to the SI unit of luminance).

stroboscopic effect
An illusion caused by oscillation in luminous flux, that makes a moving object appear as stationary or as moving in a manner different from that in which it is truly moving.

transmittance
The ratio of luminous flux transmitted by a material to the incident luminous flux.

uniform chromaticity scale (UCS) diagram
A chromaticity diagram in which co-ordinate scales are chosen with the intention of making equal intervals represent approximately equal steps of discrimination for colours of the same luminance at all parts of the diagram (see CIE Publication 15).

uniformity ratio
The ratio of the minimum illuminance to the average illuminance. In some instances, the ratio of the minimum to the maximum illuminance is quoted. The ratio usually applies to values on the working plane over the working area.

uplighter
Luminaires which direct most of the light upwards onto the ceiling or upper walls in order to illuminate the working plane by reflection.

upward light output ratio (ULOR)
The ratio of the total light output of a luminaire above the horizontal under stated practical conditions to that of the lamp or lamps under reference conditions.

utilance (U)
The proportion of luminous flux leaving the luminaires which reaches the working plane.

utilisation factor (UF)
The proportion of the luminous flux emitted by the lamps which reaches the working plane.

value
In the Munsell system, an index of the lightness of a surface ranging from 0 (black) to 10 (white). Approximately related to percentage reflectance by the relationship

$$R = V(V - 1)$$

where $R$ = reflectance (%), and $V$ = value.

vector/scalar ratio
The ratio of the magnitude of the illumination vector to the scalar illuminance.

visual acuity
The capacity for discriminating between objects which are very close together. Quantitatively, it can be expressed by the reciprocal of the angular separation in minutes of arc between two lines or points which are just separable by the eye. The expression more commonly used for an individual's visual acuity is the ratio of the distance at which the individual can read a line on a standard optician's chart to the standard distance at which a person with normal sight can read that line (e.g.

6/12 means that the individual can just read at 6 metres the line which a normally sighted person can just read at 12 metres).

*visual environment*
The environment either indoors or outdoors as seen by an observer.

*visual field*
The full extent in space of what can be seen when looking in a given direction.

*visual task*
The visual element of the work being done.

*weight*
An approximate correlate of Munsell value modified to give in conjunction with greyness, subjective equality of brightness in the various hues (from BS 5252).

*working plane*
The horizontal, vertical, or inclined plane in which the visual task lies. If no information is available, the working plane may be considered to be horizontal and at 0.7 m above the floor for offices, horizontal and 0.85 m above the floor for industry.

## CONVERSION FACTORS FOR LUMINANCE UNITS

| | cd/m$^2$ | stilb | cd/in$^2$ | apostilb | lambert | foot-lambert |
|---|---|---|---|---|---|---|
| **candela per square metre** | 1 | 0.0001 | 0.000645 | 3.14 | 0.000314 | 0.292 |
| **stilb** (cd/cm$^2$) | 10000 | 1 | 6.452 | 31416 | 3.14 | 2919 |
| **candela per square inch** | 1550 | 0.155 | 1 | 4869 | 0.487 | 452 |
| **apostilb** | 0.318 | 0.0000318 | 0.000205 | 1 | 0.0001 | 0.0929 |
| **lambert** (lm/cm$^2$) | 3183 | 0.318 | 2.054 | 10000 | 1 | 929 |
| **foot-lambert** | 3.426 | 0.0003426 | 0.00221 | 10.76 | 0.001076 | 1 |
| **To convert a value expressed in units in the first column to a unit in the top line, multiply by the appropriate factor** | | | | | | |

# BIBLIOGRAPHY

**Legislation applicable to lighting in general** (legislation applicable to specific lighting applications is listed in Part 2: Recommendations, Section 2.3.4.2, against each application)

Health and Safety at Work etc. Act, 1974
The Factories Act, 1961
The Fire Precautions Act, 1971
The Offices, Shops and Railways Premises Act, 1963
The Protection of Eyes Regulations, 1974 (amended 1975)

**British Standards and Codes of Practice relevant to lighting**

| | |
|---|---|
| BS 161 | Tungsten filament lamps for general service (batch testing) |
| BS 555 | Tungsten filament miscellaneous electric lamps |
| BS 559 | Electric Signs and high-voltage luminous-discharge-tube installations |
| BS 667 | Portable photoelectric photometers |
| BS 889 | Flameproof electric lighting fittings |
| BS 950 | Parts 1 and 2: Artificial daylight for the assessment of colour |
| BS 1308 | Concrete street lighting columns |
| BS 1853 | Parts 1 and 2: Tubular fluorescent lamps for general lighting service |
| BS 2049 | Kerosene (Paraffin) Lighting Appliances for Domestic Use |
| BS 2560 | Exit signs (internally illuminated) |
| BS 2818 | Ballasts for tubular fluorescent lamps |
| BS 2977 | Domestic lighting appliances for use with liquified petroleum gases |
| BS 3541 | Lighting fittings for general examination purposes in hospitals |
| BS 3677 | High pressure mercury vapour lamps |
| BS 3767 | Low pressure sodium vapour lamps |
| BS 3772 | Starters for fluorescent lamps |
| BS 4218 | Self luminous exit signs |
| BS 4533 | Parts 1, 2, 101 and 102: Luminaires |
| BS 4647 | Lighting sets for Christmas trees and decorative purposes for indoor use |
| BS 4683 | Parts 1-4 Electrical apparatus for explosive atmospheres |
| BS 4727 | Part 4 Glossary of terms particular to lighting and colour |
| BS 4782 | Ballasts for discharge lamps (excluding ballasts for tubular fluorescent lamps) |
| BS 4800 | Paint colours for building purposes |
| BS 4900 | Vitreous enamel colours for building purposes |
| BS 4901 | Plastics colours for building purposes |
| BS 4902 | Sheet and tile flooring colours for building purposes |
| BS 5101 | Parts 1-4 Lamp caps and holders together with gauges for the control of interchangeability and safety |
| BS 5225 | Part 1 Photometric measurements for luminaires |
| BS 5252 | Framework for colour coordination for building purposes |
| BS 5266 | Parts 1 and 3: Emergency lighting |
| BS 5345 | Parts 1-8: Code of practice for the selection, installation and maintenance of electrical apparatus for use in potentially explosive atmospheres (other than mining applications or explosive processing and manufacture) |
| BS 5371 | Standard methods of measurement of lamp temperature rise |
| BS 5394 | Radio interference limits and measurements for lighting equipment |
| BS 5489 | Code of practice for road lighting |
| BS 5490 | Classification of degrees of protection provided by enclosures |
| BS 5501 | Electrical Apparatus for potentially explosive atmospheres |
| BS 5502 | Design of buildings and structures for agriculture |
| BS 5649 | Parts 1 and 2: Lighting columns |
| BS 5717 | Transistorised ballasts for fluorescent lamps |
| BS 5971 | Parts 1 and 2: Safety and interchangeability of tungsten filament lamps for domestic and similar general lighting purposes |
| BS 5972 | Photoelectric control units for road lighting |
| BS 6012 | Heat test source (HTS) lamps for carrying out heating tests on luminaires |
| BS CP3 | Chapter 1, Part 2: Artificial lighting |
| BS CP153 | Part 1: Windows and rooflights, cleaning and safety |
| BS CP1003 | Parts 1,2 and 3: Electrical apparatus and associated equipment for use in explosive atmospheres of gas or vapour other than mining applications |
| BS CP1007 | Maintained lighting for cinemas |
| BS DD67 | Basic data for the design of buildings: sunlight |
| BS DD73 | Basic data for the design of buildings: daylight |
| BS PAS 44 | Bulkhead luminaires incorporating lamps |
| BS PAS 45 | 'Inverted' luminaires for use with lamps |

**CIBS Publications**

*Technical Memoranda*

| | |
|---|---|
| TM5 | The calculation and use of utilization factors |
| TM6 | Lighting for visual display units |

*Lighting Guides*
Building and civil engineering sites
Hostile and hazardous environments
Hospitals and health care buildings
Lecture theatres
Libraries
Museums and Art Galleries
Shipbuilding and ship repair
The outdoor environment

*Technical Reports and Monographs*

| | |
|---|---|
| No. 4 | Daytime lighting in buildings (and supplement) (To be superseded by a CIBS window design guide) |
| No. 7 | Flux distribution within a sector solid and total flux from a linear source (A.R. Bean) (Monograph) |
| No. 10 | Evaluation of discomfort glare – the IES Glare Index system for artificial lighting installations (and supplement) (To be superseded by a CIBS Technical Memorandum) |
| No. 13 | Industrial area floodlighting |
| No. 15 | The multiple criterion design method: a design method for electric lighting installations. |

**Recommendations and reports of the Commission Internationale de l'Eclairage**

| | |
|---|---|
| No. 2.2 | Colours of light signals 1975. |
| No. 13.2 | Method of measuring and specifying colour rendering of light sources 1974 |
| No. 15 | Colourimetry 1971 Supplement 1 Special metamerism index: change in illuminant 1972. Supplement 2 Recommendations on uniform colour spaces, colour difference equations, psychrometric colour terms 1978. |
| No. 17 | International Lighting Vocabulary 1970 |
| No. 18 | Principles of light measurements 1970 |
| No. 19/2 | An analytic model for describing the influence of lighting parameters on visual performance 1981 |
| No. 22 | Standardisation of luminance distribution on clear skies 1972 |
| No. 24 | Photometry of indoor type luminaires with tubular fluorescent lamps 1973 |
| No. 25 | Procedures for the measurement of luminous flux of discharge lamps and for their calibration as working standards 1973 |
| No. 29 | Guide on interior lighting 1975 |
| No. 38 | Radiometric and photometric characteristics of materials and their measurement 1977 |
| No. 39 | Surface colours for visual signalling 1978 |
| No. 40 | Calculations for interior lighting – basic method 1978 |
| No. 41 | Light as a true visual quantity: principles of measurement 1978 |
| No. 44 | Absolute methods for reflection measurements 1979 |
| No. 46 | A review of publications on properties and reflection values of material reflection standards 1979 |

**Lighting Industry Federation Factfinders**

| | |
|---|---|
| No. 2 | Dimming |
| No. 3 | Lamp guide |
| No. 4 | Energy and lighting |
| No. 5 | The benefits of certification |
| No. 6 | Hazardous area lighting |

## Books

### General

Lighting Handbook, Illuminating Engineering Society of North America, New York, 1981.

Interior Lighting Design, Lighting Industry Federation and the Electricity Council, London, 1977.

Textbook of Illuminating Engineering, J.W.T. Walsh, Pitman, London, 1947.

Light sight and work, H.C. Weston, H.K. Lewis, London, 1962.

The scientific basis of illuminating engineering, P. Moon, Dover, New York, 1961.

Interior Lighting, J.B. de Boer and D. Fischer, Philips Technical Library, Antwerp, 1978.

Lighting, D.C. Pritchard, Longman, London, 1978.

Developments in lighting – 1. J.A. Lynes (ed.) Applied Science London, 1978.

Developments in lighting – 2. D.C. Pritchard (ed.) Applied Science, London, 1982.

Lamps and lighting, M.A. Cayless & A.M. Marsden, (eds), Edward Arnold, London, 1983.

### Light

Light: physical and biological action, H.H. Seliger & W.D. McElroy, Academic Press, New York, 1965.

Daylight and its spectrum. S.T. Henderson, Hilger, Bristol, 1977.

Ultraviolet radiation, L.R. Koller, J. Wiley, New York, 1965.

Colour Science, G. Wyszecki & W.S. Stiles, J. Wiley, New York, 1967.

Photometry, J.W.T. Walsh, Constable & Co., London, 1958.

Optical radiation measurements, 1. Radiometry, 2. Colour measurement, F. Grum & C.J. Bartleson, (Eds) Academic Press, London, 1980.

### Vision

Eye and Brain, R.L. Gregory, Weidenfeld & Nicholson, 1977.

The Intelligent Eye, R.L. Gregory, Weidenfeld & Nicholson, 1971.

The Senses, H.B. Barlow & J.D. Mollon, Cambridge University Press, 1982.

The perception of light and colour, C.A. Padgham & J.E. Saunders, G. Bell & Sons, London, 1975.

Visual Perception, T.N. Cornsweet, Academic Press, New York, 1970.

Vision and Acquisition, I. Overington, Pentech Press, London, 1976.

The Psychology of Vision, The Royal Society, London, 1980.

The Ageing Eye, R.A. Weale, H.K. Lewis & Sons, London, 1963.

### Effects of lighting conditions

Human Factors in Lighting, P.R. Boyce, Applied Science, London, 1981.

The ergonomics of lighting, R.G. Hopkinson & J.B. Collins, McDonald, London, 1970.

### Lighting Equipment

Light sources, W. Elenbass, McMillans, London, 1972.

Electric discharge lamps, J.R. Weymouth, MIT Press, Cambridge, Mass. 1971.

Lamps and lighting, M.A. Cayless & A.M. Marsden, Edward Arnold, London, 1983.

Lighting fittings, performance and design, A.R. Bean & R.H. Simons, Pergamon, Oxford, 1968.

### Lighting Applications

Lighting calculations and measurements, H.A. Keitz, McMillan, London, 1971.

Handbook of Industrial Lighting, S.L. Lyons, Butterworths, London, 1981.

Exterior lighting for industry and security, S.L. Lyons, Applied Science, London, 1981.

Colour, its measurement, computation and application, G.J. Chamberlin and D.G. Chamberlin, Heyden, London, 1980.

Road Lighting, W.J.M. Van Bommel & J.B. de Boer, Philips Technical Library, Antwerp, 1980.

Principles of natural lighting, J.A. Lynes, Elsevier, London, 1968.

Daylighting, R.G. Hopkinson, P. Petherbridge & J. Longmore, Heinemann, London, 1963.

## Scientific Journals dealing predominantly with lighting

Lighting Research and Technology (published by CIBS).

The Journal of the Illuminating Engineering Society (published by The Illuminating Engineering Society of North America).

Lighting Design and Application (published by The Illuminating Engineering Society of North America).

International Lighting Review (published by Philips Limited).

Journal of Light and Visual Environment (published by The Illuminating Engineering Society of Japan).

Lighting in Australia (published by the Illuminating Engineering Societies of Australia).

Licht Forschung, (published by Pflaum-verlag KG).

Lux, (published by Association Francaise de l'Eclairage).

## Papers

### General

[1] Lighting and visual perception, P.A. Jay, Ltg. Res. & Technol., **3**, 133, 1971.

[2] Rationally recommended illumination levels, H.C. Weston, Trans. Illum. Engng. Soc. (London), **26**, 1, 1961.

[3] Criteria for recommending lighting levels, G. Yonemura, Ltg. Res. & Technol., **13**, 113, 1981.

[4] Inter-relationship of the design criteria for lighting installations, P.A. Jay, Trans. Illum. Engng. Soc. (London), **33**, 47, 1968.

[5] The need for a unified approach to interior lighting design parameters, R.I. Bell & R.K. Page, Ltg. Res., & Technol. **13**, 49, 1981.

[6] Comparison of some European interior lighting recommendations, D. Fischer, Ltg. Res. & Technol., **5**, 186, 1972.

### Task Performance

[7] The influence of illumination level on prolonged work performance, P.R. Boyce, Ltg. Res. & Technol., **2**, 74, 1970.

[8] Age, Illuminance, Visual Performance and Preference, P.R. Boyce, Ltg. Res. & Technol. **5**, 125, 1973.

[9] Illuminance, Difficulty, Complexity and Visual Performance, P.R. Boyce, Ltg. Res. & Technol., **6**, 222, 1974.

[10] Hue Discrimination and Light Sources, P.R. Boyce & R.H. Simons, Ltg. Res. & Technol., **9**, 125, 1978.

### Preferred Conditions

[11] Design of the visual field as a routine method, J.M. Waldram, Trans, Illum Engng. Soc. (London), **23**, 113, 1958.

[12] Preferred luminance distribution in windowless spaces, J.B. Collins & C.G.H. Plant, Ltg. Res. & Technol., **3**, 219, 1971.

[13] The role of the level and diversity of horizontal illumination in an appraisal of a simple office task, J.E. Saunders, Ltg. Res. & Technol. **1**, 37, 1969.

[14] Consistency and variation in preferences for office lighting, P.R. Tregenza, S.M. Romaya, S.P. Dawe, L.J. Heap & B. Tuck, Ltg. Res. & Technol., **6**, 205, 1974.

[15] Subjective judgements of lighting in lecture rooms, P. Stone, K.C. Parsons & S.D.P. Harker, Ltg. Res. & Technol., **7**, 259, 1975.

[16] Task and background lighting, A.R. Bean, & A.G. Hopkins, Ltg. Res. & Technol., **12**, 135, 1980.

[17] Users attitudes to some types of local lighting, P.R. Boyce, Ltg. Res. & Technol., **11**, 158, 1979.

### Directional effects of lighting

[18] Studies in interior lighting, J.M. Waldram, Trans. Illum. Engng. Soc., (London), **19**, 95, 1954.

[19] The flow of light into buildings, J.A. Lynes, W. Burt, G.K. Jackson and C. Cuttle, Trans. Illum. Engng. Soc. (London), **31**, 65, 1966.

[20] Lighting patterns and the flow of light, C. Cuttle, Ltg. Res. & Technol. **3**, 171, 1971.

[21] Lighting for texture, J.A. Lynes, Ltg. Res. & Technol., **11**, 67, 1979.

### Colour in Lighting

[22] A lighting engineer looks at colour, J.G. Holmes, Trans. Illum. Engng. Soc. (London), **30**, 117, 1965.

[23] Hue, greyness and weight, H.L. Gloag, Building Materials, **29**, 7, 1969.

[24] A study of user preferences for fluorescent lamp colours for daytime and nighttime lighting, A.H. Cockram, J.B. Collins & F.J. Langdon, Ltg. Res. & Technol., **2**, 249, 1970.

[25] Colour rendering tolerances in the CIE system, M.B. Halstead, D.I. Morley, D.A. Palmer, & A.G. Stainsby, Ltg. Res. & Technol., **3**, 99, 1971.

[26] Colour rendering systems and their applications, M.B. Halstead, Light & Lighting, **69**, 244, 1976.

[27] Illumination, colour rendering and visual clarity, H.E. Bellchambers & A.C. Godby, Ltg. Res. & Technol., **4**, 104, 1972.

[28] An investigation of the subjective balance between illuminances and lamp colour properties, P.R. Boyce, Ltg. Res. and Technol., **9**, 11, 1978.

29 A study of emotional reactions to light and colour in a school environment, K.T. Tikkanen, Ltg. Res. & Technol., **8**, 27, 1976.

30 The aesthetic function of colour in buildings: a critique, T.W.H. Whitfield, & T.J. Wiltshire, Ltg. Res. & Technol, **12**, 129, 1980.

*Glare*

31 The development of the IES glare index systems. Trans. Illum. Engng. Soc. (London), **27**, 9, 1962.

32 A simplified method of estimating glare index, J.C. Lowson, Light & Lighting, **55**, 291, 1962.

33 A comparison of the accuracy of methods of calculating IES glare index, P.R. Boyce, Ltg. Res. & Technol., **4**, 31, 1972.

34 The accuracy of the IES Glare Index System, L. Bedocs & R.H. Simons, Ltg. Res. & Technol., **4**, 80, 1972.

35 Individual and group differences in discomfort glare responses, P.T. Stone & S.D.P. Harker, Ltg. Res. & Technol., **5**, 41, 1973.

36 Relationship between two systems of glare limitation, H.E. Bellchambers, J.B. Collins & V.H.C. Crisp, Ltg. Res. & Technol., **7**, 106, 1975.

37 Discomfort glare from ceiling mounted luminaires, P.C. Betts & D.C. Pritchard, Ltg. Res. & Technol., **36**, 230, 1974.

38 Glare from overall diffusing ceilings, Trans. Illum. Engng. Soc. (London)**30**, 21, 1965.

39 Glare from windows: current views of the problem, P. Chauvel, J.B. Collins, R. Dogniaux & J. Longmore, Ltg. Res. & Technol., **14**, 31, 1982.

*Flicker*

40 Subjective response to a.c. and d.c. fluorescent lighting, G.W. Brundrett, I.D. Griffiths & P.R. Boyce, Ltg. Res. & Technol., **5**, 160, 1973.

41 Human sensitivity to flicker, G.W. Brundrett, Ltg. Res. & Technol., **6**, 127, 1974.

*Contrast Rendering*

42 Performance and comfort in the presence of veiling reflections, J.B. de Boer, Ltg. Res. & Technol., **9**, 179, 1978.

43 Some effects of veiling reflections in papers, J. Reitmaier, Ltg. Res. & Technol.,**11**, 204, 1979.

44 The application of contrast rendering factor to office lighting design, P.R. Boyce & A.I. Slater, Ltg. Res. & Technol. **13**, 65, 1981..

45 Office lighting for good visual task conditions, Building Research Establishment Digest 256, BRE, Watford, 1981.

46 Is equivalent sphere illuminance the future? P.R. Boyce, Ltg. Res. & Technol., **10**, 179, 1978.

*Maintenance of Lighting Systems*

47 The maintenance of lighting installations, W. Robinson, & J.W. Strange, Trans. Illum. Engng. Soc. (London), **20**, 157, 1955.

48 Economics of planned lighting maintenance, V. McNeill, Light & Lighting, **59**, 225, 1966.

*Daylight*

49 Development and practice with daylighting of buildings, J.A.M. Bell, Ltg. Res. & Technol., **5**, 173, 1973.

50 Visual aspects of sunlight in buildings, E. Ne'eman, Ltg, Res. & Technol., **6**, 159, 1974.

51 Quantitative data on daylight for illuminating engineering, J. Krochmann, & M. Seidl, Ltg. Res. & Technol., **6**, 165, 1974.

52 Sequences for daylighting design, J.A. Lynes, Ltg. Res. & Technol., **11**, 102, 1979.

53 Improved daylight data for predicting energy savings from photoelectric controls, D.R.G. Hunt, Ltg. Res. & Technol., **11**. 9, 1979.

*Emergency Lighting*

54 Emergency lighting, D. Heatlie-Jackson, et al., Light & Lighting, **65**, 226, 1972.

55 Illuminance, diversity and disability glare in emergency lighting, R.C. Simmons, Ltg. Res. & Technol., **7**, 125, 1975.

56 Conditions of emergency lighting, W. Jaschinski, Ergonomics, **25**, 363, 1982.

*Controls*

57 The light switch in buildings, V.H.C. Crisp, Ltg. Res. & Technol., **10**, 69, 1978.

58 Simple expressions for predicting energy saving from photoelectric control of buildings, D.R.G. Hunt, Ltg. Res. & Technol., **9**, 93, 1978.

59 Predicting artificial lighting use – a method based upon observed patterns of behaviour, D.R.G. Hunt, Ltg. Res. & Technol., **12**, 7, 1980.

60 The luminance distribution of an average sky, P. Littlefair, Ltg. Res. & Technol., **30**, 169, 1981.

61 Improved daylight data for predicting energy savings from photoelectric controls, D.R.G. Hunt, Ltg. Res. & Technol., **11**, 9, 1979.

62 Observations of the manual switching of lighting, P.R. Boyce, Ltg. Res. & Technol., **12**, 195, 1980.

63 Energy conservation in artifical lighting, Building Research Establishment Digest 232, BRE, Garston, 1979.

64 The energy managment of artifical lighting use, V.H.C. Crisp & G. Henderson, Ltg. Res. & Technol., **14**, 193, 1982.

*Calculations*

64 The zonal method of computing coefficients of utilization and illumination on room surfaces, J.R. Jones & J.J. Neidhart, Illum. Engng. (New York), **48**, 141, 1953.

65 The indirect component of illumination in artifically lit interiors, R.G. Hopkinson, Light & Lighting, **48**, 315, 1955.

66 Predetermination of luminance by finite difference equations, P.F. O'Brien & J.A. Howard, Illum. Engng. (New York), **54**, 209, 1959.

67 Large size perfect diffusers, H. Zijl, Philips' Technical Library, 1960.

68 Lighting design and luminance coefficients, J.R. Jones & F.K. Sampson, Illum. Engng. (New York), **61**, 221, 1966.

69 A method for the calculation of direct illuminance due to area sources of various distributions, R.I. Bell, Ltg. Res. & Technol., **5**, 99, 1973.

70 Illuminance and utilance calculation for luminous ceilings, H.E. Bellchambers & A.C. Godby, **5**, 195, 1973.

71 Yes, but which illuminance?, C. Cuttle, Light & Lighting, **66**, 349, 1973.

72 Utilance values and uniformity in a model room, A.R. Bean, Ltg. Res. & Technol., **7**, 169, 1975.

73 The calculation of utilization factors, A.R. Bean & R.I. Bell, Ltg. Res. & Technol., **8**, 200, 1976.

74 The concept of the luminaire domain, J.A. Lynes, Ltg. Res. & Technol., **7**, 185, 1975..

75 The luminaire domain and the flow of light, J.A. Lynes, Ltg. Res. & Technol., **7**, 242, 1975..

*Lighting in relation to other building services*

76 Integrated services in modern office buildings, J. Crichton & M. Wood-Robinson, Ltg. Res. & Technol., **5**, 69, 1973.

77 Development of integrated ceiling systems, L. Bedocs & M.J.H. Pinniger, Ltg. Res. & Technol., **7**, 69, 1975.

78 The installed performance of integrated ceiling systems, P.W. Smith & N. Bunker, Ltg. Res. & Technol., **13**, 58, 1981.

79 Lighting and thermal comfort, D.A. McIntyre, Ltg. Res. & Technol., **8**, 121, 1976.

# INDEX

**CIBS**

**The Chartered Institution of Building Services**

The Chartered Institution of Building Services was formed in March 1976 by the granting of a Charter to the Council of the Institution of Heating and Ventilating Engineers. The Charter made provision for the amalgamation of the IHVE (founded in 1897) and the Illuminating Engineering Society (1909), and the amalgamation took place on the first of January, 1978.

The main objects of the Institution, as specified in the Charter, are:

(1) the promotion for the benefit of the public in general of the art, science and practice of such engineering practices as are associated with the built environment, and

(2) the advancement of education and research in building services engineering, and the publication of the useful results of such research.